Analyzing Japanese High Technologies

AF851035

Analyzing Japanese High Technologies: The Techno-Paradigm Shift

Fumio Kodama

Pinter Publishers
London and New York

© Fumio Kodama 1991

First published in Great Britain in 1991
and reprinted in 1992 by
Pinter Publishers, 25 Floral Street, London WC2E 9DS

All rights reserved. No part of this publication
may be reproduced, stored in a retrieval system,
or transmitted by any other means without
the prior permission of the copyright
holder. Please direct all enquiries to the
publishers.

British Library Cataloguing in Publication Data
A CIP catalogue record for this book is available
from the British Library

ISBN 0 86187 835 3

For enquiries in North America please contact PO Box 197, Irvington, NY 10533.

Library of Congress Cataloging-in-Publication Data

Kodama, Fumio.
Analyzing Japanese high technologies : the techno-paradigm shift / by Fumio Kodama.
p. cm.
Includes bibliographical references and index.
ISBN 0-86187-835-3
1. High technology industries—Japan. 2. Technological innovations—Japan. I. Title.
HC465.H53K63 1991 91-2547
338.4'762'000952—dc20 CIP

DE MONTFORT UNIVERSITY LIBRARY
Acc. No 042549017
Date 27.10.99
Loc./Form
Class 338.4760952
KOD

Contents

List of figures

List of Tables

Preface

This book presents an empirical analysis of the generation, innovation, and diffusion of Japanese high technologies. It is not another parable of Japan's technological success.

Many people think that the Japanese socioeconomic system leads to her international competitiveness even in high technology areas. But the question remains: is this because the Japanese system differs from the Western system, or is it because a paradigm shift is occurring in technology. Put another way, is this because the Japanese built a unique socioeconomic system which is efficient in producing civilian high technologies, or is it because the existing Japanese system happens to fit the newly emerging techno-paradigm?

In order to answer this dichotomic question, first of all we have to understand what the new paradigm looks like. In this book, therefore, I have sought to formulate the characteristics of high technologies, by generalizing individual Japanese experiences. In pursuing this goal of generalization, however, I have been compelled to use the modeling approach, which is an abstraction of reality. However, I have described several concrete cases from which an abstraction is made.

Another purpose of this book is to collect my past articles published in academic and professional journals and presented at international meetings, into a book which can be read in a broader context and be available worldwide. For the purpose of synthesis, I am going to use 'techno-paradigm shift' as an intellectual integration mechanism. Although this word is difficult to define precisely, I could not find another expression which better described my thoughts.

Ideal readers are assumed to possess an academic background in mathematical modeling, though the arguments can be understood even if the mathematical details are not studied in depth. Professional experience in science policy will also prove helpful, though non-experts will benefit from reading this book, because policy implications are described in a broader context. This book is meant for several groups of readers, those conducting research in the field of policy science including graduate students, and managers of in-house training programs in government and industry.

This book comes out through my collaboration with my graduate students at Graduate School of Policy Science in Saitama University. I am particularly indebted to Yukichi Honda, Masahiko Kobayashi and Toru Muraoka, who produced excellent master theses under my guidance. In integrating their individual works into a book, I owe very much to generous support given by NISTEP (National Institute of Science and Technology Policy). Many people including visitors at NISTEP helped me with this integration. Particularly, I would like to express my thanks to Diana Hicks, a visiting fellow from SPRU, for the tremendous efforts she made in reading and giving useful suggestions on my draft.

My deepest gratitude goes to my wife, Minako, and my three children, without whose patience and unwavering support, it would have been impossible to sustain my research for so many years.

Fumio Kodama
September, 1990, Tokyo

List of acronyms

AI	Artificial Intelligence
BEA	Bureau of Economic Analysis
BTL	Bell Telephone Laboratories
CAT	computerized axial tomography
C&C	computer and communication
CGW	Corning Glass Works
CMOS	complementary metal oxide semiconductors
CRT	cathode ray tube
DOD	Department of Defence
DRAM	dynamic random access memory
DSM	dynamic scattering mode
E&E	energy and electronics
ENNA	Engineering Advancement Association of Japan
ERA	Engineering Research Association
ETL	Electro-Technical Laboratory
FMS	flexible manufacturing system
IC	integrated circuit
IEA	International Energy Agency
IEEE	Institute of Electrical and Electronics Engineers
M&A	mergers and acquisitions
MC	machining centre
MCVD	modified chemical vapor deposition
ME	microelectronics
MIT	Massachusetts Institute of Technology
MITI	Ministry for International Trade and Industry
MOS	metal oxide semiconductors
MTBF	mean time between failures
NC	numerically controlled
NEC	Nippon Electric Company
NIES	newly industrializing economies
NISTEP	National Institute of Science and Technology Policy
NSG	Nippon Sheet Glass
NSK	Nippon Seiko
NTT	Nippon Telegraph and Telephone Corporation
OECD	Organization for Economic Cooperation and Development

PATTERN	planning assistance through technical evaluation of relevance numbers
PERT	program evaluation and review technique
PIPS	pattern information processing system
QC	quality control
RCA	Radio Corporation of America
R&D	research and development
SAPPHO	scientific activity predictor from patterns with heuristic origin
S/Es	scientists and engineers
SEI	Sumitomo Electric Industries
SIC	Standard Industrial Classification Code
SITC	Standard International Trade Classification
SPRU	Science Policy Research Unit
S&T	science and technology
UPSO	US Patent and Trademark Office
VAD	vapor phase axial deposition
VLSI	very large scale integration
VTR	video tape recorder

Chapter 1

Overview: six categories of techno-paradigm shift

When geographical shifts occur in the balance of world industrial strength, those countries overtaken often attack the newcomers with accusations of a 'conspiracy'. In many cases this conspiracy is actually the spontaneous innovation of institutions by the newcomers, and these innovations in turn become the world model for a new industrial era.

As these new innovations fall outside the experience of the previous powers, they indeed appear to be some sort of conspiracy. As the targets of this criticism are unaware that something new is being done, they cannot effectively refute the charges against them.

This phenomenon is described as a 'historical paradox': a less developed country, in the process of trying to catch up with the advanced countries, unconsciously resorts to innovative policies that are even more progressive and modern than those of the more advanced countries, and only later begins to appreciate why it has chosen these policies. This sort of 'historical paradox' was witnessed at the turn of the century with England and France on one side and Germany on the other.[1] It also now seems to have appeared between Japan and the nations of the West.[2]

Many specialists have been pointing out changes in the basic pattern of technological innovation.[3] With the emergence of 'high-technology', various changes are occurring in the whole framework of science and technology policy, whether it be governmental policy or corporate policy. These changes are significant enough to merit the label: 'paradigm shift'.[3]–[6]

This shift is making obsolete the policy arguments of science and technology which have hitherto been common sense in theories of business administration and international relations. Because of the lack of full appreciation of the paradigm shift in science and technology, these phenomena, and malfunctions such as mismatch in management practices, the paradox in economic policy and international disputes, are occurring.

Such changes are everywhere: who makes high-technology available; how it is generated; and what it is utilized for. They are in the field of manufacturing companies, and in their principal business: the economic actors by whom high-technology is brought into the market. They are in research and development (R&D) activities, and in technology development processes: the human intellectual activities which generate high-technology. They are also in the innovation pattern, and in technology diffusion: the societal process through which high-technology is realized.

First, a fundamental redefinition of the manufacturing company is taking place. The manufacturing company is traditionally a site for production and the economist's formulation is a production function: capital plus labor make things. But in many Japanese manufacturing companies, R&D investment is much *greater* than capital investment. R&D investment surpassed capital investment quite recently and the change occurred rapidly. This signals a paradigm shift; if R&D investment begins to surpass capital investment the corporation could be said to be shifting from being a place for *production* to being a place for *thinking*.

Second, there are changes in business. In the past, one technology used to correspond to company. But now, especially in Japan, technological diversification has progressed so much that it is hard to distinguish a company's principal business from its secondary business. In many cases the principal business of a company is now overtaken by its secondary business.

Corporate diversification in the United States is mostly the result of mergers and acquisitions (M&A). According to several analyses there, corporate growth through diversification is surprisingly low, and many attempts at diversification have even ended in failure. However, US analyses of diversification through M&A cannot explain the Japanese situation. Today's leading Japanese firms have entered the stage where they survive by adapting to the environment, relying on consistent, dependable R&D.

Third, major changes are observed in the field of research investment decision-making in industry. Investment decisions are no longer based on rates of return. It is more like the principle of surf-riding: the waves of innovations come one after another and you have to invest; if you miss you are killed. The pattern of competition is also changing; the competitor used to be another company within the same industrial sector, but in many cases nowadays the competitor is a company in a different industrial sector.

As the challenge for high-tech leadership could come from seemingly unrelated industries, without regard to country of origin, international agreements among companies in the same industrial sector to avoid protectionist pressure could easily be rendered meaningless. Conversely, companies could form technological alliances across national and industrial boundaries, competing for development with each other. High technology may thus change the conventional wisdom, the common sense in theories of business administration and international relations.

Fourth, there are changes in the technology development process. In the high-tech era, the key issue of technology policy has become not how to break through technological bottlenecks, but how to put existing technology to the best possible use. Accordingly, a day of reckoning has come for technology policy, which traditionally has emphasized the supply side of technology development. A need has now arisen for a technology policy which works from the demand side.

In developing new policies to meet this need, the most important element is the process of 'demand articulation'. Through this process, the need for a specific technology manifests itself and the R&D effort is targeted toward developing and perfecting it.

The public policy implication can be found in the Japanese government-sponsored research consortia. One which caught particular international attention was the VLSI (very large scale integration) research association which included all five of Japan's IC (integrated circuit) chip manufacturers at the time. In this research association, rather than focusing on the method of production itself, research efforts emphasized developing a prototype for IC manufacturing equipment and analyzing a process for the crystallization of silicon. In other words, potential users of optical steppers and of silicon materials joined together to articulate their needs.

In the United States, pre-competitive research is usually carried out at a university under the sponsorship of several private corporations. This represents a chronological concept of technological innovation, in which research begins at the scientific stage and progresses through the application and development stages. In Japan, however, pre-competitive research achieved through research associations is better represented by plotting industrial linkages on a graph of coordinates, in which the goal is to create an engineering infrastructure as the basis for competition. The key point to bear in mind is that there exist two separate types of pre-competitive research.

Fifth, there are the changes in innovation patterns. Conventional wisdom holds that technical innovation is achieved by breaking through the boundaries of existing technology. With regard to recent innovations in new fields such as mechatronics and optoelectronics, however, it would be more appropriate to view innovation as fusing different types of technology rather than as technical breakthroughs.

There are two types of technical innovation. One is the traditional 'technical breakthrough', of which the transistor offers a typical example. This appears to be the forte of the West. Another is 'technology fusion', a typical example of which would be the mechatronics revolution. This type of innovation is Japan's strong point.

Our analysis will clearly suggest that the technical innovations involved in high technology are more of the fusion type than of the breakthrough type. Expressed more accurately, a single technical breakthrough alone is not sufficient for progress in high technology. Rather, only through the organic fusing of several technical breakthroughs in a number of different fields can a new technology be created.

The *last* change is the technology diffusion that reflects a shift from technical change to institutional inertia. According to Christopher Freeman, the widespread generalization of information technology, not only in the 'leading' branches but also in many branches of the economy, is possible only after a period of change and adaptation of many social institutions to the potential of the new technology. Whereas technological change is often very rapid, there is usually a great deal of inertia in social institutions.

The diffusion of facsimile communication in Japan was not initiated before the revision of the Public Telecommunication Law. On the other hand, the diffusion of CAT (computerized axial tomography) scanners in Japan is a case where the relevancy of this technology to institutional changes accelerated rather than discouraged acceptance. If a technology is very novel to users and has an intrinsic usefulness to them, as occurred with CAT scanners, the introduction of the new technology triggers a chain reaction of institutional changes which creates a favorable environment for the rapid diffusion of the new technology.

The quantitative analysis of computer diffusion in the forty-seven Japanese prefectural governments reveals that its diffusion process is not as simple as described so far in terms of technoeconomic paradigm changes. Our study yields two levels of dichotomy. One dichotomy is between organizational complexity and technical complexity. Organizational inertia is in fact a determinant of the diffusion of computer utilization when computers are applied to those activities which are complex organizationally. The other dichotomy is between technologies applied to old activities and those applied to new activities. Institutional inertia is not critical to the diffusion of computers applied to new activities for which the institutional framework is not yet fully developed.

These six categories of paradigm shift are summarized in Table 1.1. All categories are derived from my past studies published in various academic and professional journals. Before going into detail, however, I am going to highlight these categories of techno-paradigm shift.

Table 1.1 Six categories of techno-paradigm shift

1. Manufacturing companies:
   from *producing* to *thinking organization*
2. Business dynamics:
   from *single technology* to *technology diversification*
3. R&D activities:
   from *visible competitors* to *invisible enemies*
4. Technology development:
   from *linear progression* to *demand articulation*
5. Innovation patterns:
   from *technical breakthrough* to *technology fusion*
6. Technology diffusion:
   from *technical change* to *institutional inertia*

1. Shift in manufacturing companies

Manufacturing companies are the primary participants in technological innovation; the leading factor in the various changes in this field is ongoing change in the fundamental definition of the corporation. The traditional archetype of a manufacturing company is a group of people who produce high-quality products at the lowest possible cost, using the most advanced equipment. According to recent statistics, however, R&D expenses account for more than equipment investment at many high-tech companies.

Among the top fifty Japanese companies that had the largest R&D expenditure in 1987, the R&D expenditures in as many as twenty-seven out of these fifty companies, exceeded spending for capital investment. For example, in the top company, Toyota Motor Corporation, R&D expenditure was 275 billion yen while its capital investment was Y225 billion. The Matsushita Company spent ¥271 billion on R&D, but invested only ¥66 billion in capital equipment. Hitachi spent ¥262 billion on R&D, and ¥103 billion on capital investment.

If you take the national total of all the Japanese manufacturing companies, R&D expenditure has already surpassed the capital investment: ¥6.10 trillion on R&D expenditure and ¥4.86 trillion on capital investment. Thus, the R&D/capital investment ratio was 1.26:1.0. In other words, the Japanese manufacturing companies were spending 26 per cent more on R&D than on equipment.

What is more important is that this change occurred only recently. The time-series data of the national total show that the reversal took place only recently in 1986. The manufacturing firms' ratio of R&D to capital investment was 0.62 in 1980, 0.91 in 1985 and 1.17 in 1986. However, a closer look at these trends reveals that the reversal is attained by the opposite movements of these two expenditures: the capital investment is constantly decreasing, while the R&D investment is ever growing. Subsequently, these two curves crossed in 1986.

Various forms of 'soft automation' such as the flexible manufacturing system (FMS), reduced industry's need for capital expenditure sharply. Such flexible machinery is merely reprogrammed, rather than replaced, when a factory is being switched over to making a new product. FMS evolved as an aid to the transition in production methods from the mass production of a small product range to the small-batch production of a diversified range of products. Thus a comparison of these two production systems would help clarify what exactly FMS is.

We have a saying in Japan which might have been the motto of a 'wild west' gunfighter. Namely that the technology of mass production requires that every 100 shots fired must hit the same target bulls-eye, whereas the technology of FMS requires that every single shot fired should hit the target bulls-eye. The key difference lies in the fact in the former cases we are talking about all 100 shots hitting the same bulls-eye, whereas in the latter case each shot should hit the target, but the target may change with each shot.

When looked at from the viewpoint of technical backgrounds, the technological capability level of FMS should be higher than that of mass production by a factor of one or two, in every aspect from process accuracy, quality control, reliability, maintenance, through worker skill requirements. Another way of putting this is to say that FMS can exist only when both machines and men are capable of a very much higher technical level than in a mere mass production system.

Therefore, we might paradoxically say, the introduction of FMS has brought the situation in which manufacturing companies are getting rid of the burdens related to manufacturing. Thus, it has become possible for manufacturing companies to increase R&D investment which is related to the future.

When R&D expenses are greater than capital investment in many companies, today's leading Japanese firms have entered the stage where they survive by adapting to the situation, based on consistent, dependable R&D. The corporate archetype is, therefore, changing from 'producer organization' to 'thinking organization'.

2. Shift in business dynamics

Technological innovation drives structural changes in corporations. The Management and Coordination Agency of Japan monitors corporate R&D trends in its annual report, *Survey on Research and Development*. The section on 'corporate research expenses by industry and product' in this report has been better organized every year since 1970. The agency asks companies to reveal their R&D investment in thirty-two product categories. The collected data gives an idea of how R&D expenses are distributed by product category in each of twenty-one industries. Therefore, an analysis of diversification can be made using this database.

According to 1980 data for the textile, metal manufacturing and shipbuilding industries, R&D investment was smaller in major product categories than in minor ones. The ratio of R&D investment in areas outside principal product fields to total R&D expenditures in 1980 was 0.55 in the textile industry, 0.53 in metallic products and 0.67 in shipbuilding. In 1986, the figures were 0.71, 0.55 and 0.79, respectively, indicating increasing diversification. Side businesses are becoming mainstays in terms of R&D expenditure and an index of future corporate activity.

For example, the textile manufacturing industry underwent a major structural change in the 1970s, and several of its side businesses are gaining in importance. Product categories with no R&D investment in 1970 that have since gained a large share of total corporate R&D investment are construction materials and medical/pharmaceutical products. One textile company is using its fiber technology to produce kidney dialysis equipment and construction materials, maintaining the technological continuity.

Using the database, we can measure the degree of diversification of R&D investment in various product categories and make an industry-by-industry comparison. The industry that saw the most drastic diversification in the 1970s was textiles; that with the least was the automobile industry; and the steel industry was somewhere in between. The data indicate that the greater the sales growth for the main products, the smaller and slower the diversification of the industry. On the other hand, those industries that are experiencing saturated demand for their main products are well ahead in diversification plans.

However, a recent survey reveals that the Japanese steel industry has finally changed since the beginning of 1980s. In the 1970s the percentage of R&D expenses in the product fields apart from steel stood at about 20 per cent of their total R&D expenditure during 1970s. In 1986 it reached almost 50 per cent of the total. In other words, only half of the steel industry remains in the iron and steel, as far as the R&D activities are concerned.

The textile and shipbuilding industries are diversifying 'upstream', faced with tough international competition due to the strong showing by Asian NIEs (newly industrialized economies). This type of diversification strategy is intended to offset decreased exports of finished products with increased exports of manufacturing machinery and equipment.

Diversification 'downstream' is notable in the electric machinery industry including communications equipment/electronics, and in the precision machinery industry. These two ranked first and second in terms of growth in the 1970s, therefore the industrial growth of the 1970s was realized in the form of expanded applications for precision and electronics technologies, which characterized the recent mechatronics revolution.

The pharmaceutical industry has been actively diversifying both upstream and downstream. This has much to do with the clear direction for technological development set by the emergence of biotechnology and other major technologies, which are unlike the traditional chemical technologies.

3. Shift in R&D activities

High technology developments render traditional management obsolete. The microelectronics industry presents a typical innovation pattern in a high-tech area. For example, vacuum tubes were replaced by transistors, which then gave way to IC's (integrated circuits). Eventually, VLSI circuits superseded ICs. In other words, new devices were invented and adopted one after another, like waves crashing onto the beach.

The case of DRAM (dynamic random access memory) chips shows how fast, and sometimes destructive, innovations can be. In this area, new products have been introduced to the market almost every three years since 1972. One-kilobit DRAMs were put on the market in 1972, 4-kilobit in 1975, 16-kilobit in 1978, 64-kilobit in 1981 and 256-kilobit in 1983.

What is characteristic in this technological innovation race is that manufacturers introduce new products before the learning process is complete on the technology of the preceding innovations. Competition may be characterized as predatory: new devices drive their predecessors completely out of the market within six years of introduction.

With this kind of technological advance, corporate decisions on investment are not made on the conventional basis of the rate of return; they are made according to the principle of surf-riding. Companies have no choice but to invest to ride the wave of innovation, or they are left behind by competitors. Investment must continue in order just to stay in the market.

Cases like these in the microelectronic industry greatly concern competitors in technological innovation in the same industry. However, solutions to basic technical problems have sometimes been found in areas outside the conventional industry. Therefore, in many cases the competitor is now a company belonging to a different industrial sector.

For example, in the Japanese development of optical fibers, the first production was by a glass manufacturer. The original product had two problems – mechanical fragility and high transmission loss – which were not solved by the glass-maker. To reduce transmission loss, a glass-maker naturally thought in terms of purifying the glass, which obviously has a physical limit. This problem was solved by NTT (Nippon Telegraph and Telephone Company), a user of optical fibers, by discovering that using longer wavelengths drastically reduces the transmission loss. The fragility was solved by a cable manufacturer who invented a coating technology that reduced fragility. Thus those two problems were solved by companies in other industrial sectors. The mass-production method for optical fibers was developed by a consortium of the cable manufacturer and NTT, not by the glass-maker.

Basic high-tech problems are occasionally solved by ideas or technologies born in totally unrelated fields. Thus high-tech companies have to monitor not only direct competitors in their own sector but also firms in other industries. In effect, this means that high-tech companies must engage in R&D competition with 'invisible enemies'.

Given this, we can imagine an innovation model under which leaders in technological advances replace their predecessors regardless of industry, and revolutionary innovation coincides with a change of leadership. According to such a model, the change of leadership could occur across national boundaries, thus making it impossible to define international competition in technological development using conventional patterns.

4. Shift in technology development

Demand articulation is essential to technology development activity. From the view point of scientists and technologists, demand articulation is the search and selection process among technical options. However, the sample population from which technical options are to be drawn varies over a wide spectrum of sources of innovation. It may be drawn from the shelf of existing technical collections. It may also be drawn from the pool of scientific knowledge, as is advocated by scientists and technologists in terms of a linear progression model of innovation.

Between these two extremes there is a wide range of technology development in which some parts may be drawn from existing technologies, while some parts may be drawn from the scientific pools or sometimes beyond them. In fact, this range of technology development may cover the majority of technology development.

In the development of liquid crystal technology, demand articulation for more advanced, sophisticated products helped foster innovative new technologies. There were clear commercial applications for liquid crystal displays, and the companies which were most successful in this field were those whose technological goals were most suitable to such applications.

Watch manufacturer Suwa Seiko, whose overriding objective was the development of an electronic quartz watch, had specific needs for technology which would extend the life of its product, reduce energy consumption and allow for a smaller, thinner display. At the same time, the Sharp Company was caught up in stiff competition in the development of pocket calculators. Their key to success was creating smaller, thinner products which used little energy. For example, it would be bothersome to carry around a calculator whose batteries had to be replaced after every ten hours of use.

When liquid crystals were invented, the standard for display technology was the cathode ray tube (CRT). RCA (Radio Corporation of America), which had itself invented the liquid crystal display, chose to stick with the CRT technology, as did most manufacturers of CRT screens.

What turned out to be the correct technological choice was made by the more specialized manufacturers. They limited their applications to displays for calculators and watches, which had a specific need to reduce energy requirements. In other words, they were the only manufacturers who achieved demand articulation *vis-à-vis* liquid crystal display technology.

In the development of Japanese language word processors, it was a cultural factor that made demand articulation possible. The Japanese language uses a mixture of the 50-character *kana* alphabet and over 3,000 Chinese *kanji* characters. The 'keyboard' of a traditional Japanese typewriter – actually a plate printed with rows of tiny characters, and a mechanical arm with a pointer to select them – contains over 3,000 characters which the operator must find and punch in one by one. 'Typists' were highly trained specialists.

Manufacturers were forced to return to fundamental research, starting from the basic principles of linguistics. After ten years of wide-ranging research at Toshiba's laboratories, a new method was finally devised, in which *kana* could be converted to *kanji*. The result was the birth of a new product concept: the Japanese language word processor. Working from a keyboard of forty-eight *kana*, the operator can simply key in text as it is pronounced, while a computer programmed with a dictionary of Japanese vocabulary and grammar semi-automatically converts the *kana* letters into *kanji* characters where necessary.

One of public policy implications of demand articulation can be found in the rich Japanese experiences in research consortia, in which rival firms co-operate. Over seventy government-sponsored research consortia have been formed in Japan so far. One which caught particular international attention was the VLSI research association which was in existence from 1976 to 1979. The association included all five of Japan's IC chip manufacturers at the time: NEC, Toshiba, Hitachi, Fujitsu and Mitsubishi Electric.

Cooperation among rival firms requires that the subject for research is both fundamental and of common interest to all participants. Rather than focusing on the method of production itself, research efforts emphasized the development of a prototype for IC manufacturing equipment and analyzing a process for the crystallization of silicon, a basic material in chip production. In addition, no manufacturers of either production equipment or IC chip materials were among the participants. The inclusion of such firms would invariably have resulted in a conflict of interest among the participants.

The specific activities of the VLSI association included the development of an optical stepper. The five companies joined together to establish a joint research facility, whose lithography division enlisted camera manufacturers Canon and Nikon to create an optical stepper. At the same time, the results of research into the silicon crystallization process were passed to The Shin-Etsu Semiconductor Company and The Osaka Titanium Company. These two companies formed Japan's silicon production industry.

In the days before the VLSI research association was established, Japan relied on imports of production equipment and silicon crystals. From the viewpoint of technology policy, the research association provided the domestic engineering infrastructure required for IC chip production. Competition among the association's participants focused on the skillful utilization of the new equipment and materials, and their successful integration into the group's own production lines. It was accepted that a domestic manufacturer capable of producing high performance equipment and materials which met the needs of users would be beneficial to all IC chip manufacturers.

The establishment of the VLSI research association, therefore, can be interpreted as a policy mechanism in which the government, working in cooperation with IC chip producers, bore the costs needed to set up the engineering infrastructure for IC chip production.

It is generally said that, in R&D directed at new products, equal importance is assigned to knowing what to make and finding out how to make it. In the VLSI research association, potential users of optical steppers and silicon materials joined together to articulate their needs. The costs associated with this represented half those for the R&D, and if it is taken into account that this burden was shared by the government and the users of that technology, this could be justified as a public policy.

5. Shift in innovation pattern

'Technology fusion' yields innovations from breakthroughs. There are two types of technical innovation. One is the traditional 'technical breakthrough', of which the transistor offers a typical example. Another is 'technology fusion', a typical example of which would be the mechatronics revolution.

A number of revealing comparisons can be made between technical innovations achieved through technology fusion and those achieved through technical breakthroughs.

First, while technical breakthroughs become possible when a prominent corporation in a specific industry takes a leadership role, technology fusion is made possible by joint operations among related industries. The mechatronics revolution in Japanese machine tools first became possible through cooperation between Fanuc, a spin-off of a communications equipment manufacturer which developed the controller, NSK, a bearing manufacturer – which developed the ball screw – and the materials manufacturer which developed the teflon material used to coat the sliding bed.

Second, while technical innovations of the breakthrough type bring rapid growth for a particular corporation, technology fusion contributes to gradual growth in all the industries concerned. As a result of the semiconductor revolution in the United States, for example, many previous manufacturers of vacuum tubes went bankrupt while new corporations such as Texas Instruments experienced rapid growth. In contrast to this, the mechatronics revolution brought growth to existing firms in Japan's machine tool industry.

Third, while technical breakthroughs are promoted through defense policies, technology fusion is promoted through industrial policy. Although the mechatronics revolution occurred in 1975, as early as 1971 machinery and electronics-related laws, which had been created independently, were combined and enacted as the Law for the Development of Specified Machinery and Electronics Industries. This law utilized the phrase 'consolidation of machinery and electronics into one', indicating a clear intention to fuse machinery and electronics technology.

The process of innovation through technology fusion begins with R&D investment by a certain corporation in a product field outside its primary product lines. Unilateral investments in research in a particular industry are no more than efforts at diversification which do not necessarily result in technical innovations.

In those cases in which two industries engage in 'reciprocal' research investment, however, there is the possibility that a new field of technology will be formed. For example, the ceramics industry had long been investing in research into electrical equipment, but it was not until the electrical industry undertook investment in ceramics research that new ceramics emerged as a technical field.

Technology fusion involving reciprocity in research investment can be analyzed both quantitatively and systematically. The most prominent fields of high technology, such as mechatronics, new ceramics and biotechnology, can be considered products of technology fusion. The mechatronics revolution emerged in 1975 through reciprocal research investment in the four fields of ordinary machinery, precision instruments, electrical machinery and communications/electronics. Biotechnology appeared in 1974 through reciprocal investments between the fields of food, pharmaceuticals and industrial chemicals, while new ceramics appeared in 1982 through investments across the fields of ceramics, ordinary machinery and electrical machinery.

It could be argued that this hypothesis on reciprocal research investment is more than simply an operational definition for the analysis of technology fusion. In other words, if technology fusion is simply the mixing together of different types of technology, it would be possible simply through the merger of corporations possessing different types of technology. Such a combination of different technologies through mergers, however, would merely involve one technology making use of another, with little likelihood of bringing about the technology fusion discussed here. On the contrary, in reciprocal investment which takes the form of two corporations undertaking joint research, the two technologies would be of equal importance and technology fusion could easily result. The hypothesis of reciprocity is thus intrinsic in technology fusion itself.

The fact that the 'new materials revolution' is identified as a product of technology fusion reflects the current situation in which expectations regarding advances in materials technology are greater now than ever before. Materials have clearly played an important role throughout history. How, then, are today's expectations of materials technology different from those of the past?

A recent survey by Japan's MITI (Ministry for International Trade and Industry) drew up a list of major fields of technology, and asked the participants from each field to list those fields, outside their own, from which they are most eagerly anticipating new advances. It revealed great expectations regarding new materials and low expectations regarding aerospace and energy.

It is widely known that state-sponsored aerospace and energy projects, many of which are military-related, have produced numerous breakthroughs in new materials, in the form of spinoffs. The low expectations regarding aerospace and energy, however, seem to indicate that the development of new materials is generally no longer expected to follow the conventional mechanisms of the past.

New materials are now expected to be developed in direct response to specific technological needs, rather than to emerge in a roundabout way from defense-related projects. Increasing expectations regarding new, directly-applicable materials technology reflects, at least in part, a dramatic change in the nature of technological innovation in the materials field.

The terms 'materials design' and 'custom materials' have also recently entered the scientific lexicon. This means that molecular structures can now be artificially manipulated to produce specific material for specific uses. It is also hoped that, in the future, biological processes will come to replace chemical processes in the manufacture of materials.

6. Shift in technology diffusion

British economist Christopher Freeman points out four categories of technical change: incremental innovations, radical innovations, change of technology system and changes in the techno-economic paradigm. The diffusion of steam power and of electric power are obvious examples of the last category. So too, is the combination of innovation associated with the electronic computer. He argues that the widespread generalization of a new paradigm is possible only after a period of change and adaptation of many social institutions to the potential of the new technology.

However, there exists a surprisingly small amount of empirical and quantitative evidence for such a techno-economic paradigm change, although such changes are mentioned by several authors in general ways. Therefore, empirical and quantitative studies are reported here on the diffusion process of techno-economic paradigm changes.

For descriptive evidence, we conducted two case studies in the fields of telecommunication and welfare, because these two fields are most affected by regulation. One example is the progress being made in developing new technologies for facsimile communications in Japan. Given the particular character of the Japanese language, potential demand for this type of communication has surfaced, taking advantage of the liberalization of the use of telephone circuit networks through the revision of the telecommunication law.

The Revision of the Public Telecommunications Law in 1971 allowed the telephone network to be used for purposes other than voice transmission, and enabled facsimile equipment to be connected to subscriber lines. This meant that charges could be levied according to the length of time and distance used for communication, and not at a fixed rate. The market that had hitherto been limited to special applications and specific businesses was now wide open to general business, and research and development in both terminal and network technologies was launched.

In the field of welfare-oriented products, the technology has to be diffused within the framework of the existing national welfare system. However, the introduction of new technology sometimes induces the changes in regulation. The development of CAT scanners is an example.

The first CAT scanner was developed by EMI of the United Kingdom, but its diffusion rate was low in Japan. When it was decided that Queen Elizabeth would visit Japan, the Japanese government decided to purchase thirty-three CAT scanners. These scanners were distributed to brain surgery departments in major public hospitals.

When the number of CAT scanners surpassed 200 units, a professional association of experts was established. This association demanded that the Ministry of Health should permit the use of scanners to be covered by the national health insurance scheme. The government decided to do so, and then afterwards the market for scanners expanded rapidly.

For quantitative evidence, the diffusion of computer systems into the local governmental sector is analyzed, because it is obviously not a leading sector as far as computer utilization is concerned. Moreover, a unique Japanese database is available, which can accommodate the requirements for quantitative study of the organizational aspects which affect diffusion of a new techno-economic system.

The quantitative analysis reveals that the diffusion process is not as simple as described so far in terms of techno-economic paradigm changes. First, a distinction between the installation of the computer and its utilization must be made. The installation behavior follows conventional wisdom: the psychological approach is valid, the diffusion phenomena is epidemic. However, when it comes to utilization behavior, the theory of techno-economic paradigm change holds: institutional inertia is a determinant of diffusion rather than technological sophistication.

However, several caveats apply to the simple argument posed by the techno-economic paradigm change. Our study yields the two levels of dichotomy. One dichotomy is one between organizational complexity and technical complexity. Organizational inertia is in fact a determinant to the diffusion of computer utilization in those activities which are complex organizationally. However, organizational inertia is not a determinant of diffusion when the computer is utilized for those activities which are less complex organizationally but which require technical sophistication.

The other dichotomy is whether a new technology is applied to old activities or it is applied to new activities. Institutional inertia is a determinant when computers are applied to conventional activities for which the institutional framework is already well established. On the other hand, institutional inertia is not critical to the diffusion of computers when they are applied to a new activity for which the institutional framework is not yet fully developed, i.e. when the institutional framework is to be developed in parallel with the diffusion of the new technology.

7. Some comments on research methodology

Roughly speaking, there are two methodologies used in studying technological innovation: one is the 'case study' approach which is used by scholars in business school; and, the other is the 'mathematical modeling' approach which is used by economists. However, both approaches seem to have intrinsic deficiencies.

The case study approach may generate insight and sometimes the insight can be surprising, but the findings are likely to be specific to the object and to the environment of the case study. And policy implications can be drawn out mainly through analogy. By comparing several cases, we might structure past experiences, and thus we might learn important lessons. However, if the whole paradigm is changing, the past experiences will be little use to decision-makers who are mainly interested in the future.

Moreover, case study data is collected through interviews with the people who developed the technology. In many cases, it appears that irrational decisions have been made because the engineers cannot relate the bounded rationality behind these decisions. They might have done something very innovative that they themselves are not aware of. This is very true when paradigms are shifting. Usually, it is very difficult for people to recognize their innovative accomplishments.

Using the modeling approach, on the other hand, we can generalize our past experiences. However, when it comes to the study of technology, we have an intrinsic dilemma: while science is universal, technology is specific in every aspect. Therefore, risk of over-generalization is always involved in mathematical modeling. Understandably, technologists complain about these approaches, saying that it is difficult to identify what kinds of technologies model builders are talking about.

Another dilemma in the modeling approach is that a rigid conceptual framework is necessary for modeling, and available frameworks are very often derived from past observable phenomena. Although established frameworks are the most appropriate for analyzing past technologies, when a new paradigm of technology is emerging, there might be a serious discrepancy between the subject of study and the methodology used for the study. We have to create a new conceptual framework for a new paradigm. In fact, Schumpeter succeeded in making a breakthrough in the study of technology by characterizing the innovation process as a 'creative gale of destruction' of the existing order, in which we might include existing frameworks for analysis of technology.

Therefore, we might conclude, case study without generalization is nothing but storytelling, while a generalized approach without specific cases is anything other than a study of technology. In the following chapters, I will try to strike a balance between these two conflicting approaches. This balance could be attained by taking a new approach to both the case studies and the mathematical modeling. However, it can be most effectively realized by bringing the mathematical modeling closer to the case studies rather than the reverse. I will try to approach it from the 'policy science' perspective, the area in which I spent the past twenty years of my academic career.

The 'policy science' approach is essentially a kind of 'engineering approach' in the sense that the aim is to obtain meaningful policy implications even if the approach might be piecemeal. The method is to develop analytical tools which are most appropriate for a specific problem and to apply them to available databases. In essence, this is a scientific endeavor associated with a pragmatic art: it deals with dynamic aspects of the technology: it departs from an academic obsession with a static comprehensive framework hoping for the elimination of all kinds of inconsistency. In terms of technology policy analysis, it aims at acquiring an effective understanding of the technology within an appropriate level of generality rather than attempting an unobtainable unified theory of technology.

However, our recent historical experiences show that the policy science approach is easy to describe but difficult to implement. This ideal was first advocated by scholars almost twenty years ago [7], but very few successful realizations have been produced. As a result, disillusionment with the possibility of implementing the ideal is prevalent today in scientific communities throughout the world. In order to regain lost interest, a few successful studies which manifest this ideal of policy science analysis are needed.

In order to demonstrate the possibility of this type of analysis, I have done my best to organize this book in the following manner. Every chapter starts with a few specific case studies relevant to the subject. These will serve to establish the basic mechanisms in each category of techno-paradigm shift. Then, efforts are directed toward identifying the most appropriate database for generalizing those basic mechanisms discovered in cases studies. Thereafter, a mathematical formulation of the problem is described. As you will see, the majority of mathematical modeling was done by the author, but sometimes efforts were devoted to choosing the most appropriate from among existing models. Finally, each chapter is concluded by policy implications in which either retrospective evaluation of past policies in a new context or future directions for policy research are described.

Throughout this book, I tried to follow this structure as far as possible. However, the specific objects studied in case descriptions are not necessarily the same as the objects dealt with by quantitative analysis. These discrepancies arise mainly from the limited availability of databases and/or of analytical tools. In fact, filling these gaps might serve to formulate an agenda for future research.

References

[1] P. Valéry, 'La conquéte allemande', *The New Review* (London), vol. 16, no. 92, January 1897, and 'Une conquéte méthodique', in *Collection Les Amis d'Edouard*, Champion, 1924.

[2] K. Nishio, 'What is the modern Japan?' (in Japanese), *Chyuou-Kouron*, December 1982, pp. 216–36 (Chyuou-Kouron-Shya, Tokyo).

[3] C. Freeman, *Technology Policy and Economic Performance*, pp. 60–79, Pinter Publishers, 1987.

[4] F. Kodama, 'The six changes in techno-paradigm' (in Japanese), *The Journal of the Institute of Electronic, Information and Communication Engineers*, vol. 73, no. 1, pp. 27–30, 1990.

[5] F. Kodama, 'Major changes in innovation process in Japan', *Les Nouvelles (Journal of the Licensing Executives Society)*, forthcoming.

[6] F. Kodama, 'Tidal changes underlying R&D and technology development' (in Japanese), in *General Survey on Research Development and Technology Development*, F. Kodama (ed.), Sangyo-Chosakai, 1989, pp. 1–5.

[7] E. Quade, *Analysis for Public Decisions*, Elsevier Science Publishing, 1975.

Chapter 2
Manufacturing companies and business dynamics: from producing to thinking organization for diversification

1. Introduction

It is widely accepted that restructuring a country's economy requires technological changes. Recently, industrial research and development has been identified as one important source of technological change. However, there is surprisingly little solid quantitative evidence linking industrial R&D with economic restructuring.

Japanese companies are now investing more in R&D than in plant and equipment. High-tech firms are spending on average 80 per cent more on R&D than plant and equipment. When R&D expenses are greater than capital investment at many companies, we need to rethink what manufacturing companies are. Today's leading Japanese firms have entered the stage where they survive by adapting to the changing environment, based on consistent, dependable R&D. The corporate archetype is, therefore, changing from *producing* organization to *thinking* organization. However, why is this possible?

One reason is that factory robots and other forms of soft automation have been installed, reducing industry's need for capital expenditure sharply. Such flexible machinery is merely reprogrammed, rather than replaced, when a factory is being switched over to making a new product. The changeover is completed in hours instead of weeks; the cost of doing so is marginal.

A second reason is that several world crises have endangered Japan's low-tech manufacturing. To survive, many manufacturing firms have had to abandon much of their old core businesses and hasten where possible into higher-value markets. In the process, they are having to add all sorts of high-tech components to their products.

From the viewpoint of national policy, however, a drastic change in a country's economic structure occurs when a new industry is created, an existing industry disappears, or both. Therefore, in this connection a more relevant dichotomy is whether or not a sector's industrial R&D is done within or outside its principal product fields. The industrial R&D by each sector within its principal product fields is supposed to be directed toward keeping existing industries viable. The industrial R&D activity outside its own principal product fields is directed toward creating new industries.

Each sector's R&D activity outside its principal product fields can be considered technological diversification of that industry. Therefore, a study of technological diversification of existing industries leads to the study of creation of new industries. This is true in Japan where internal corporate venturing [1][2] is a dominant mode for creating new industries based on technological diversification.

In this chapter, we will describe analyses of the first two categories of techno-paradigm shift described in the preceding chapter: a shift in manufacturing companies, from producing organizations to thinking organizations; and a shift in business dynamics, from single-based technology to technological diversification.

First, we will present a recent change occurring in the Japanese manufacturing company in terms of its R&D investment and its capital investment. Analysis will reveal that the reversal is taking the form of the sharp reduction in industrial capital expenditure, while production continues to increase. The R&D expenditure, on the other hand, does continue to grow.

In the *second* part of this chapter, therefore, our analysis is concentrated on technical innovation in manufacturing technology. The introduction of FMS has paradoxically brought about the situation in which we will not have to worry about manufacturing anymore; thus the increase in R&D investment is made possible. Accordingly, we will describe the unique features of FMS.

In the *third* part of this chapter, to start with an approach to measuring an industry's technological diversification behavior is described. Then, the sectoral patterns of diversification behavior in Japanese industry are identified. Finally, Japanese corporate and governmental policies relevant to diversification are discussed in terms of their mechanisms and effectiveness.

2. Changing pattern of manufacturing companies

The traditional archetype of a manufacturing company is a group of people who produce high-quality products at the lowest possible cost, using the most advanced equipment. According to recent statistics, however, R&D expenses account for more than equipment investment at many high-tech companies.

Among the fifty companies that had the largest R&D expenditure in 1987, R&D expenditure in many of them exceeded the spending for capital investment. In as many as twenty-seven out of the fifty companies, R&D investment began to surpass capital investment. In order to avoid the bias which might be brought about by yearly fluctuations in annual capital investment, each company's recent three-year average from 1985 through 1987 is calculated, and is compared with its three-year average R&D expenditure during the same period, as shown in Table 2.1.

Table 2.1 R&D expenditure compared with capital investment in major Japanese companies (average in 1985–7)

Company	R&D expenditure (¥ million) (A)	Capital investment (¥ million) (B)	A/B
1 Toyota Motor	258,333	301,333	0.86
2 Hitachi	251,773	123,367	2.04
3 NEC	235,667	180,667	1.30
4 Matsushita Electric	204,647	41,971	4.88
5 Toshiba	179,133	127,433	1.41
* NTT	178,897	1,674,388	0.11
6 Fujitsu	163,637	113,833	1.44
7 Nissan Motor	160,000	106,533	1.50
8 Honda	114,867	80,380	1.43
9 Mitsubishi Electric	113,000	77,000	1.47
10 Sony	92,978	77,968	1.19
11 Mitsubishi Heavy Industry	87,333	65,022	1.34
12 Mazda Motor	77,433	102,667	0.75
13 Nippondenso	59,860	86,233	0.69
14 Canon	59,563	53,579	1.11
15 Sharp	59,164	56,176	1.05
16 Nippon Steel	55,000	140,000	0.39
* Tokyo Electric Power	47,282	1,118,844	0.04
17 Fuji Photo Film	43,278	42,533	1.02
18 Sanyo Electric	41,945	56,261	0.75
19 Ricoh	36,720	26,671	1.38
20 Takeda Chemical Industries	34,900	22,440	1.56
21 Kobe Steel	34,233	71,865	0.48
22 Asahi Chemical Industry	33,833	52,033	0.65
23 Kawasaki Steel	31,267	100,133	0.31
24 Ishikawajima-Harima	31,267	12,933	2.42
25 Bridgestone	29,667	36,467	0.81
26 Sumitomo Metal Industries	28,671	79,133	0.36
27 Mitsubishi Kasei	28,000	31,200	0.90
28 Matsushita Communication	26,437	4,365	6.06
29 NKK	26,405	85,377	0.31
30 Fuji Heavy Industries	25,867	50,180	0.52
31 Matsushita Electric Works	25,200	24,480	1.03
32 Sumitomo Chemical	24,500	35,479	0.69
33 Oki Electric Industry	24,467	32,167	0.76
* Kansai Electric Power	24,139	587,000	0.04

(continued)

Table 2.1 (Continued)

Company	R&D expenditure (¥ million) (A)	Capital investment (¥ million) (B)	A/B
34 Komatsu	23,233	14,433	1.61
35 Fuji Electric	22,698	18,038	1.26
36 Asahi Glass	21,000	45,667	0.46
37 Kubota	20,952	24,391	0.86
38 Kawasaki Heavy Industries	20,432	23,333	0.88
39 Sankyo	20,367	12,933	1.57
40 Victor Co. of Japan	20,326	29,187	1.01
41 Sumitomo Electric Industries	20,167	27,833	0.72
42 Fujisawa Pharmaceutical	19,754	7,219	2.74
43 Eisai	19,382	5,927	3.27
44 Shionogi	19,191	8,786	2.18
45 Toray	18,100	41,850	0.43
46 Omron Tateishi Electronics	18,086	20,442	0.89
47 Konica	17,870	15,804	1.13
48 Aisin Seiki	17,145	23,389	0.73
49 Mitsui Toatsu Chemicals	16,295	22,536	0.72
50 Hino Motors	15,967	12,536	1.25
* Chubu Electric Power	15,562	488,334	0.03

Note: The companies with * are non-manufacturing.

Source: *Japanese Companies Handbook.*

In the Table 2.1, the top fifty companies are listed in decreasing order of R&D expenditure. In the last column of the table, the ratio of R&D expenditure to capital investment is calculated for each company. These ratios, for example, are 2.04 for Hitachi, and 1.30 for NEC. In other words, these companies are spending 104 per cent or 30 per cent more on R&D than on plant and equipment.

The listing of companies by R&D expenditure inevitably includes companies such as NTT (Nippon Telegraph and Telephone Corporation) and The Tokyo Electric Power Company, that primarily provide the service. They are denoted by * in the left column of the table. In these non-manufacturing companies, capital expenditure is much larger than R&D expenditure and thus the R&D/capital investment ratio is not larger than 0.10. If these non-manufacturing companies are excluded, the average R&D/capital ratio of the top fifty Japanese manufacturing companies is 1.29. In other words, major Japanese manufacturing companies are spending on average 29 per cent more on R&D than on equipment.

However, what is more important is that this change occurred only recently. The time-series data in the national totals of all the Japanese manufacturing companies are shown in Table 2.2.

Table 2.2 Time-series of R&D expenditure compared with capital investment of all Japanese manufacturing companies ¥ billion

Year	R&D expenditure	Capital investment	R&D/cap. ratio
1980	2,896	4,651	0.62
1981	3,374	5,161	0.65
1982	3,756	5,099	0.74
1983	4,257	4,762	0.89
1984	4,777	5,788	0.83
1985	5,544	6,110	0.91
1986	5,740	4,896	1.17
1987	6,101	4,860	1.26

Source: MITI for capital investment, and Prime Minister's Office for R&D expenditure.

As shown in the table, the reversal had taken place already in 1986. In 1986, the amounts were 5.74 trillion yen for R&D expenditure and 4.90 trillion yen for capital investment; in 1987, they were ¥6.10 trillion and ¥4.86 trillion, respectively. The ratio of R&D to capital investment was 0.62 in 1980, 0.89 in 1983 and 0.91 in 1985. However, the ratio became 1.17 in 1986, and reached as high as 1.26 in 1987, meaning that Japanese manufacturing companies were spending 26 per cent more on R&D than equipment.

We are interested in the time trend of the distribution of these ratios for individual companies because the phenomena of R&D expenditure surpassing capital investment has meaning when it is referred to individual companies, although not so much in the case of the total of all the manufacturing companies. We are also interested in how general this phenomenon is throughout the major Japanese manufacturing companies, not only in the R&D intensive companies. It might be argued that the choice of fifty companies in terms of their R&D expenditures might bias the results in favor of R&D expense *vis-à-vis* capital investment. Therefore, the total sales of companies was selected as a criterion for choosing the fifty manufacturing companies, in order to eliminate the biases and to be neutral. Thus the time shift in cumulative distribution curves of the R&D/capital investment ratios among the largest fifty manufacturing companies with the largest total sales, is shown in Figure 2.1.

In this figure, if the curve shifts toward right and down, R&D is exceeding capital investment in more companies. As is depicted in the figure, a trend in that direction is obvious. In other words, the reversal is

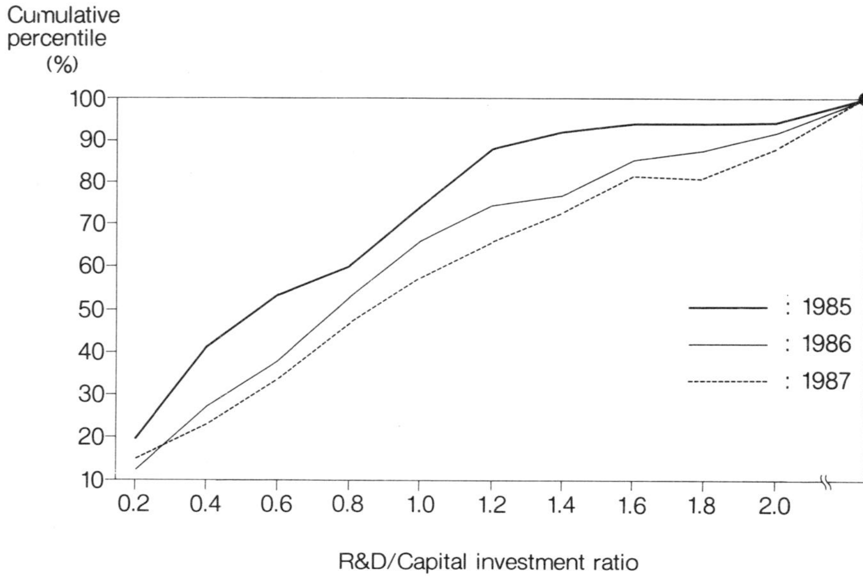

Figure 2.1. Cumulative distribution curves of R&D/capital investment ratios among the largest fifty manufacturing companies. *Note*: The largest fifty manufacturing companies are selected based on their total sales in each year (Source *Japanese Companies Handbook*)

getting more and more common in Japanese manufacturing companies. In 1987, the curve looks almost linear, meaning that the ratio is almost uniformly distributed among the fifty companies.

For international comparison, the corresponding US data of the average of 1978–82 is compared with the Japanese average data of 1984–7, as shown in Table 2.3. The time period of the US data is much older than that for the Japanese data, and the US data only covers companies in the chemicals, pharmaceuticals, electronics, electrical machinery and automobile sectors. However, it is hard to believe that if recent data had been included it would have shifted the cumulative curve downward. Therefore, in spite of this inconsistency, the comparison indicates clearly that R&D investment surpassed capital investment in only a minority of the US firms. Therefore, this phenomenon is unique to Japanese manufacturing companies.

A closer look at Table 2.2 reveals, however, that the reversal is attained by the opposite movements of these two expenditures: capital investment decreased continuously since 1985, while R&D investment has grown since 1980; then the two curves cross in 1986. Therefore, we first should ask why capital investment is decreasing.

As described before, various forms of 'soft automation' such as FMS, reduced sharply industry's need for capital expenditure. The

Table 2.3 United States–Japan comparison of R&D/capital ratios among the fifty manufacturing companies

Range of R&D/capital ratio	Percentile of manfacturing companies	
	Japan	USA
⩽ 0.0	0%	0%
⩽ 0.2	13	4
⩽ 0.4	30	36
⩽ 0.6	39	60
⩽ 0.8	54	78
⩽ 1.0	67	92
⩽ 1.2	76	98
⩽ 1.4	85	98
⩽ 1.6	93	100
⩽ 1.8	96	100
⩽ 2.0	96	100
⩽ 5.0	100	100

Source: for US data, Standard & Poors Inc., 'Computer statistics service'.

introduction of FMS has created the situation in which manufacturing companies will not have to worry about manufacturing anymore. Thus the development of manufacturing technologies made the increase in R&D investment possible. The corporate archetype is, therefore, changing from a producing organization to a thinking organization.

3. Flexible manufacturing system

FMS is regarded as a method of production which permits a diversified product range. This type of small-batch production of a diversified product range has been common in the clothing business for a long time. However, the processing of metal products is an entirely different and much more difficult proposition. It is, therefore, fair to say that the term FMS became common only after it was applied to the metal processing industry enabling small-batch production of a wide range of metal products.

There seem to have been four factors present in the background to this development: (1) the development of higher precision in mechanical technology; (2) the electronic revolution represented by microprocessors; (3) the emergence of robot technology; and, (4) the perfecting of quality control technology. In other words, although FMS had existed for quite some time, its mass scale application only really took off once these four factors had made their contribution.

When analyzed from the market side, the development of FMS seems to be inextricably connected with fierce intercompany competition and the saturated state of capital goods' and consumer goods' markets. For these two reasons, manufacturing companies inevitably seek to give their products competitive edge by a point of distinction. A company cannot hope to survive in the face of a saturated market and ferocious competition if it plans to produce a product that is merely similar to those of its competitors or merely the standardized product of average level: such a strategy will merely lead to either just the maintenance or loss of market share without contributing to the overall growth of the market.

The reason why FMS has been so successfully applied in Japan is that Japanese industry urgently needed and was looking for a means of small-batch production of a diversified range of products. As is known, Japanese industry has a dual structure with a small number of giant companies and a very large number of small and medium size companies which are highly interrelated. The giant companies are able to take advantages of the economies of mass production of a quality product. Smaller companies, on the other hand, are tied to producing a wider range of products in small batches as they are locked in by tough contracts to the production systems of large companies, to whom reliability of delivery at a fixed price is vital to the survival of small firms. Thus there is, in the strict sense, a subcontractor relationship, but in another sense, almost a division of labor. It is, therefore, readily understandable that such characteristics of the Japanese industrial structure should both have required and nurtured the development of FMS to the point where Japan is the market in which it is most widely used.

Hence, in this section, FMS will be considered from such viewpoints as technical background, its impact on employment and the cultural background in Japan where it was originated. Finally, the future trend of FMS and its impact on mass production will be considered. ([3]–[4])

3.1 Technical background of FMS

As a catch-phrase FMS seems simple and understandable, but it is always difficult to establish what its technical nature is. Among the various ways to clarify this point, one is to consider why FMS was regarded as a necessity. FMS enables a transition in production methods from the mass production of a small product range to the small-batch production of a diversified range of products. Thus a comparison of these two production systems would help clarify exactly what FMS is. This comparison will be made in five parts: (1) process accuracy, (2) quality control, (3) reliability, (4) maintenance and (5) worker skill requirements.

Process accuracy

As described before the technology of mass production requires that every 100 shots fired must hit the same target, whereas the technology of FMS requires that every single shot fired should hit the target.

In the case of a mass-production system, one can expect both a shake down period for the machines, and a concurrent familiarization for the worker to the machine as production experience increases. It can also be expected that once all the bugs have been ironed out of a single product mass-production system, certain latent production potential which is greater than expected may be discovered. In the case of FMS, however, this cannot be expected.

When looked at from the viewpoint of process accuracy, the technological capability level of FMS should be higher than that of the mass production by a factor of one or two. Another way of putting this is to say that FMS can exist only when both machines and men are capable of a very much higher level of process accuracy than in a mere mass-production system.

Quality control

The techniques of quality control (QC) were first seen in the United States whose principle is to statistically inspect the product at the final station of the production line. This technique was revised in Japan so as to permit checks on the quality at each station of the production line and this modification was a spectacular success, e.g. in the automobile industry and others. This improvement is, however, only within the quality control technique for the mass-production system.

The basic idea of QC in FMS is that the machines automatically stop when a product is found to be non-standard/out of specification. You can better understand this by supposing an unmanned factory at night. Suppose you find that a machine in the unmanned plant has turned out a whole series of substandard products, unnoticed by anyone. How does this happen and how can it be prevented?

In the QC procedure as applied in mass-production systems, the inspection is made after the processing is finished, irrespective of whether the inspection is carried out at the final station or at each intermediate station. In the case of QC as applied in FMS in which each individual product may be of different specification, the inspection should take place based on signals received before processing. For example, obviously a tool bit should be changed before it becomes worn out and causes reject products, and so this is done by measuring the torque against preset standards.

Maintenance

The widespread belief that Japanese QC is a kind of preventive maintenance because it is carried out at each station by the operators, is not correct since it is merely based on signals/observations after a product has been processed. In FMS there is a true technology of preventive maintenance in that it is implemented before a product has been processed

The maintenance of a FMS, which is expected to give most benefit to the management in terms of unmanned operation at night, requires a completely different concept of management from the mass-production system in which the attainment of production quality is the main object.

In other words, it is important to minimize the down-time of the operation in the case of daytime operations involving the workers. Conversely in the case of FMS, it is important in unmanned operation at night not to amass a heap of reject products; it is preferable that the entire system should immediately stop when a substandard product is found, and wait for maintenance workers to arrive whilst the system remains stopped. The key factor for FMS, therefore, lies in attaining such maintenability that the system stops the moment a substandard or reject product is found in the production line.

Reliability

A plant in which all the products may be individually different from one another and produced by unmanned operation, when viewed from the standpoint of reliability would seem to be almost the ultimate technological achievement. In other words, the technical reliability of FMS is a factor of one greater than that of mass production.

The Kofu Plant of the FANUC company, where this technological new frontier is being opened up, has an output per employee of 100 million yen per year. Trial calculations as to whether the plant could reduce its costs by 10 per cent by improving plant/equipment reliability, gave the startling result that this would result in equipment trouble as rarely as once every four years. This extremely high reliability is not yet common everywhere.

However, this level of reliability is not yet attained in general. MTBF (mean time between failures) is some way from achieving the level of unmanned operation. [5] This shows that the widespread operation of unmanned factories cannot be attained if it is to be based on existing technology only. To attain this goal, such new technology as fault detection or self-checking functions need to be developed.

Skill requirement

At FANUC's fully automated Kofu Plant, the only tasks for workers are the loading of a workpiece onto the guideway to the machining center (MC) by the use of a hoist, and the preparation of computer programs for the numerically controlled (NC) units. In other words, worker input is in the form of two separate kinds of human labor. The manual, physical labor of using the hoist and the intellectual, cerebral work of programming.

In such mass-production systems as the automobile industry, there are people who describe the Japanese workers as being multi-skill whereas American or European workers are single-skill. The level of the multi-skill involved in Japan is merely that a workman will either carry out various types of operations such as welding, screwing and making electrical connections, or sequentially operate a number of machines; all eventually fall within the definition of manual work.

What the workers at the FANUC plant are involved in, however, is a combination of both physical and cerebral activity. In the FANUC case, multi-skill is not confined to manual work. Thus FMS can be regarded as a unique technical system with the underlying assumption that the skill requirement of workers must differ from that of workers in the mass-production system; compared with the usual multi-skill concept, the requirement is a factor of one greater.

3.2 Impacts on employment problems

The reason that microelectronics (ME) is discussed in relation to employment is that it is essentially different from conventional automation technology.

In conventional automation technology there was a kind of paradox; namely that it was found that those jobs which required high degrees of skill or accuracy, jobs which manual workers found difficult, were easier to automate, whilst a job which workers found easy were more difficult to automate. Put another way, the machines preferred tasks requiring power, speed, accuracy, continuity, repetition, monotony, etc., and humans preferred the opposite kind of work which they found easier. This paradox produced the division of labor between workers and machines. However, at the same time, the excessive division of labor alienated humans from machines in the production environment.

Now let us consider what a robot is capable of. This representative of mechatronics does not fit the stereotype of what conventional machines are good at. It is not that powerful. It is not that fast moving. Its working accuracy is also lower than machine tools by a factor of two. But it has the ability to carry out complicated tasks and, especially when married to computers, it can execute a variety of tasks with selectivity. From this point of view, a robot behaves with more of the characteristics of a man than a machine. Paradoxically enough, this similarity is now making it easier for workers and machines to coexist.

The background to this character is, from the viewpoint of mechanical technology, that a robot has a high degree of freedom in movement, whereas the degree of freedom of an ordinary machine is usually limited to one axis. In other words, an ordinary machine is mono-functional because its degree of freedom is limited. In contrast, it is designed to have higher rigidity and strength, to operate at higher speed and at higher accuracy. In this way, a robot is a rather unique product from the viewpoint of mechanical technology. [6]

JAPANESE EXPERIENCES

In 1980, the Ministry of Labor conducted a study of the impact of the introduction of NC/MC machine tools and transfer machines tools and transfer machines on employment. 2,000 manufacturers of general machinery and equipments who employed more than 30 regular employees were the target of the survey.

During the period of the survey, 1977–80, the number of places of employment which introduced this kind of equipment was almost half of the total (47.1 per cent). This period coincides with the employment adjustment period following the second oil crisis, when the number of full-time employees decreased as a whole. It should, however, be noted that the companies who introduced NC units reported an 2.5 per cent average rate of decrease in number of employees, while those who did not, reported 4.4 per cent. Moreover, among the companies introducing NC, 48 per cent increased the number of full-time employees, whereas 42.3 per cent reduced the number of full-time employees. For the companies not introducing the units, figures were 42.7 per cent and 47.5 per cent respectively.

The reasons given for this increase are: increase in orders, 68.1 per cent; expansion of business, 31.5 per cent; and increase in numbers of new jobs due to introduction of new technology, 29.8 per cent. The ways in which companies responded to the introduction of NC may be categorized as: workers were to operate the NC machines, 65 per cent; workers were transferred to other departments, 25 per cent; recruited new workers for the new job, 24 per cent; and only 4 per cent made an employment adjustment.

According to past Japanese experience, mechatronics clearly contributes to labor-saving on the one hand, whilst on the other hand having the effect of often creating a new job. [7] Thus employment will either be maintained, or at the very least, the impact on employment will be reduced by the combined effect of product expansion and order increase as a result of the launch of new products.

New problem formulation

Just as the Luddite movement did not succeed during industrial revolution in England, so the spread of robots is irreversible because of the practical benefits they offer. The problem of the relationship between technical innovation and employment should, therefore, be considered not as a question of how this innovation will be stopped, but as one of how employment should be reviewed in the light of progress in technology.

This is an anecdote set in the plant of a Japanese integrated electronics equipment manufacturer. A visitor from France asked 'Where did the workers go when automated equipment was introduced to replace them?' An executive answered with a serious look, that 'They got married!' The point of this is that the ME revolution has not had such an impact on the employment situation. The first reason for this is that economic growth has been high. Investment in labor-saving equipment was used as a means to expand production without increasing the labor requirement. The second reason is that the age of workers was low enough to enable retraining and comparatively easy redeployment.

Ultimately, there seems to be an optimum employment structure for each respective era in technology. When a new technology opens up and develops a new job for some people, the reality is that some other workers must make a transition either in their job or in the place of employment. There is a considerable time lag before the employment structure settles down in its optimum shape. The problems of technical innovation versus employment should, therefore, be defined as how to cope with this time lag.

It is never constructive to try to stop technical innovation, and ultimately it is inevitable. The question then becomes how to minimize the negative impact of ME innovation on employment. What becomes necessary is a balance between technical innovation and organizational innovation.

If product innovation based on ME innovation is actively developed and consequently if this leads to the production of new products, the influence on employment would be positive. Moreover, if FMS can meet the diversified demand from consumers and can expand overall consumption, the negative influence on employment would be reduced. Through organizational innovation society could offer workers in the old types of job destined to be superseded by ME a good chance to adapt to new types of job which might be created by mechatronics.

3.3 Cultural backgrounds of FMS

It is traditional for a diversity of products to be available in Japan. In a way, Japanese culture is a cultural system dealing in many and various products. The language system may be taken as a starting point. The Japanese language consists of 2 phonetic alphabets each composed of 50 symbols and more than 2,000 Chinese characters which are essentially ideographs. This combination of phonetic and ideographic characters is quite unique in the world both in terms of the sheer number of symbols and the number of possible combinations thereof.

Such a complicated system of written language was a significant handicap at the time when such primitive technology as mass-production systems were dominant. The recent emergence of sophisticated technology such as FMS is changing the scene in that the cultural complication serves as a stimulus to technical innovation. If one takes the example of ordinary computer equipment, the Japanese level of technology was not especially high in the days of the line-printer, the product of mass-production technology; a more flexible type of product, the dot-matrix printer raised the Japanese technological level up to the highest in the world. The dot-matrix printer is suited to the Japanese cultural background, if the target of coping with Chinese characters is to be achieved.

The traditions from the history of the Japanese manufacturing industry seem to have worked in favor of FMS. The Edo period (Edo is the former name of Tokyo before the Meiji Restoration) was one period of isolation from the fifteenth to the seventeenth centuries, during which a unique and flourishing merchant culture was nurtured by the comparatively high concentration of the population in cities. A system of city-based economics flourished wherein commerce and industry cohabited as complementary neighbors. Such an economic system will eventually require that production should be tailored to the needs of the client, the simple concept underlying small-batch production of a diversified range of products.

The main customers of FANUC are small and medium-size companies. The basic unit in the Kofu Plant consists of three components, machining center, robot and transfer system of the workpiece, and is called a 'cell'. This cell is deliberately comprised of components which are most appropriate for the average level of small and medium-size companies. By visiting the FANUC plant, the owners or technical staff from such companies can easily understand how FMS could be applied to their own factories.

In other words, FANUC uses their plant as a showroom for the FMS, and as the means by which they 'sell' the 'cell' to the visitors. During the Edo period, customers queued up for craftsmen who in turn responded to the order of each individual client by making a variety of products, often requiring the use of many different types of tools, in the presence of the customer. Such a scene is in fact an Edo-period *Ukiyoe* (wood-block print) of the FANUC cell system.

3.4 Future impact on mass production

Hitherto, FMS has been developed mainly for machining operations and is already a matured technology. The future direction of development is, however, expected to follow a shift toward assembly operation. This shift of emphasis from machining to assembly will alter the behavior of the FMS manufacturers. [5]

Nowadays you can see many manufacturers of electric household appliances and electronic equipment trying to enter this field by spending money for R&D and commercial activities. Many of them are giants whose overall technological strength will strongly influence the FMS sector of the industry. Hitherto, they have been the big client and users of FMS and thus the unique experience and technology they have experienced as users will become available to other industrial sectors.

Finally, it will be worth watching how the maturing technology of machining FMS will be fed back into mass-production technology. It has already been mentioned that the FMS technology, which aims at the diversified range of products with small-batch production, is superior to that of the mass-production system by a factor of one in terms of processing accuracy, quality control, maintenance, reliability and skill requirement. Mass-production systems struggled to attain this level of technology only to find the goal unattainable. This could indicate that mass-production technology could benefit if FMS technology is fed back into it, thereby improving its overall technological level. Obviously, for example, dot-matrix printer technology capable of printing more than 2,000 characters will naturally enable the printing of only 26 characters with higher quality and lower cost.

4. Database for diversification studies

In the preceding section, it is shown how the introduction of FMS has produced the situation in which manufacturing companies are getting rid of the burdens related to manufacturing. Thus it has become possible for manufacturing companies to increase R&D investment which is related to the future.

The recent growth in R&D investment is presumably attributable to increased investment in long-term research, and increased product development costs for secondary businesses. In other words, the recent rapid increase in R&D investment comes from the two factors: an increase in depth, toward more fundamental research; and an increase in width, toward more encompassing research.

However, in what follows, we will focus on the widening scope of R&D investment. In other words, we will concentrate our analysis on technological diversification of Japanese industry. A Japanese database on research and development that disaggregates an industry's intramural expenditure into thirty-one different product fields is used to measure technological diversification. Thus the direction of technological diversification identified using this database can be interpreted as the future metamorphosis of each industry and the sectoral pattern of industrial transformation.[8]

4.1 Japanese database

Two attempts have been made to identify a sectoral pattern of technological diversification: the work on significant British innovations by Pavitt, [9] and the work on US patents by Scherer.[10] The unusually rich Japanese R&D data source collected in the *Survey of Research and Development* produced by the statistics bureau of the prime minister's office [10] makes it possible for us to take a third approach. For all the Japanese companies with a capital of 100 million yen or more (3,803 companies in 1982), intramural expenditure on R&D is disaggregated into 31 different product fields, as shown in Table 2.4.

A company like, for instance, Hitachi was asked for this survey to break down its R&D expenditures into such categories as chemical products, fabricated metal products, ordinary machines, household electric equipments, communication and electronic equipments, automobiles, precision instruments and so forth. This is an alternative to reporting expenditures in one category assigned to Hitachi's primary industry, electrical machinery manufacturing. In the case of an expense that is difficult to classify by the kind of product, it is divided proportionally on the basis of the number of researchers.[11]

This survey is available every year since 1970, so that a time-series analysis is possible and investigations at any particular point in time can be checked up in terms of their particularity. Moreover, because this survey is collected under the special law on statistics (*Shitei-Toukei*), companies are instructed to fill in the form with a great care.

It is generally argued that R&D expenditure is nothing more than one of the input indicators for technology diversification.[12] However, compared to the R&D expenditures within the principal product fields, a

Table 2.4 Classification of principal product fields for diversification studies

Industrial sector	Principal product fields
agriculture	agricultural, forest and fishing
mining	mining
construction	building construction and civil engineering
food	food
textile mill products	textiles
pulp and paper products	pulp and paper
printing and publishing	printing and publishing
chemical products	chemical fertilizers and inorganic and organic chemical products; chemical fibers; oil and paints; other chemicals
drugs and medicines	drugs and medicines
petroleum and coal products	petroleum
rubber products	rubber
ceramics	ceramics
iron and steel	iron and steel
nonferrous metals and products	nonferrous metals
fabricated metal product	fabricated metals
ordinary machinery	ordinary machinery
electrical machinery	household electrical appliances; communication and electronic equipment;other electric equipment
motor vehicles	automobiles
other transport equipment	ships; aircraft; railroad equipment; other transportation equipment
precision equipment	precision instruments
other manufacturing	other manufacturing products

sector's R&D expenditure outside its principal product fields can be assumed to reflect some results of technological diversification. These investments can be easily canceled if the prospects are not favorable, because at that stage they are investments outside the main business activity. Such cancellation, however, is not so easy in the case of investment within the principal product fields. On this basis, a sector's R&D expenditure in a product field outside its principal product fields can be assumed to be the output indicator of technology diversification, because the increase in expense is to be expected following success in the earlier stages.

4.2 Profile of diversification

From this database, a profile showing how each industry diversified its R&D activities during the 1970s can be drawn. The textile industry was chosen as an example because it was one of the most diversified industries during this period.

This kind of profile (Figure 2.2) shows product fields as a function of the industry's R&D expenses (on a logarithmic scale). Product fields are arranged by decreasing order of expense in 1970: textile products, the largest expense item, is at the top, chemical fibers, the second largest expense item, is next, and so on. The 1970 profile is a monotone curve, whereas those for 1975 and for 1980, are no longer simple monotone curves, showing visually how the rank order of expense in product fields has changed since 1970.

The industry's expense in the product field of building construction, for example, was nothing in 1970, was substantial in 1975, and became one of the largest by 1980. There was no R&D investment in drugs and medicines in 1970, but it was the fourth largest expense item in 1980.

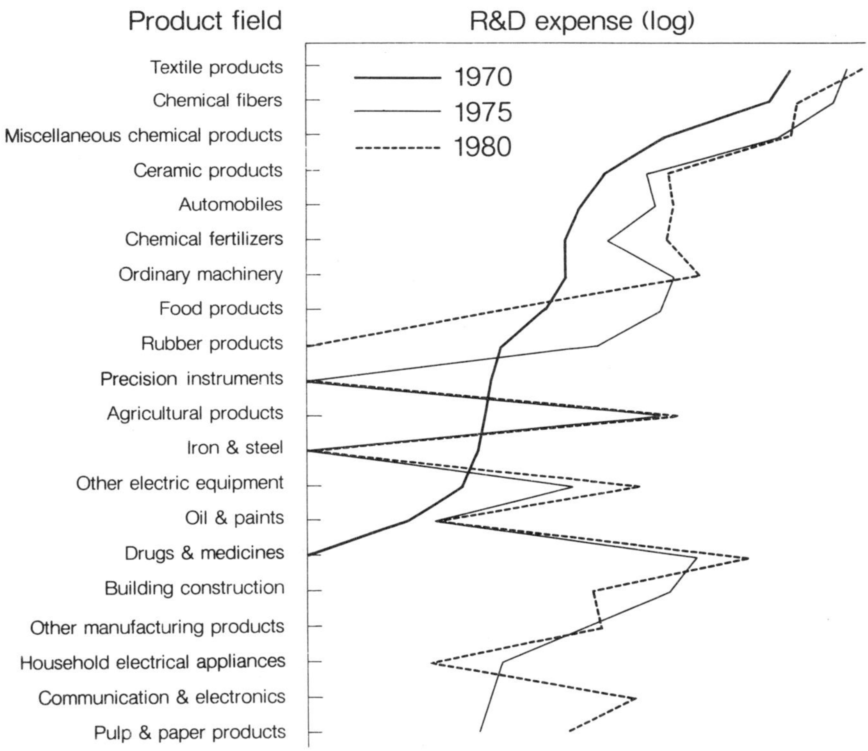

Figure 2.2. Profile chart for diversification in the textile industry

Since this profile was drawn based on R&D expenses, it does not necessarily describe the product diversification of a specific company. However, a company that was in the textile industry is now applying its fiber technology to building materials, which contain strong fibers, or to filters for medical equipment such as those used in kidney dialysis. In this way, technological continuity lies behind product diversification.

5. Measurement of diversification

5.1 Degree of diversification

Several quantitative analyses on diversification can be made from our database. First is the degree of diversification. If an industry's R&D distribution over the different product fields (they are shown in the right hand column of Table 2.4) can be supposed to be a probability distribution, we can construct an indicator for degree of diversification, using the concept of entropy.
Given

E_{ij} = *i-th* industry's R&D expense into *j-th* product field.

Let
p_{ij} = the share of *j-th* product field in the total R&D expense of *i-th* industry, i.e.,

$$p_{ij} = E_{ij}/\Sigma_j E_{ij},$$

then, H_i, the entropy of *i-th* industry, can be calculated by

$$H_i = - \Sigma_j \, p_{ij} \cdot \log_2 p_{ij}.$$

If an industry's diversification is advanced and its R&D expenses cover many product fields, with less concentration on specific fields, then the industry's entropy value, H_i, rises. On the other hand, if an industry's diversification is not as advanced and its R&D expenses cover only a few product fields with a high concentration on specific fields, then the industry's entropy value drops.

A time-series of entropy values of major industries (Figure 2.3) shows that the textile industry is the most diversified, motor vehicle manufacturing is the least diversified industry, and that iron and steel is in between. During the 1970s, the textile industry and petroleum and coal products manufacturing saw rapid diversification, while the other industries did not experience much change in diversification.

There may be several ways to interpret these results. One clear result, however, is that an industry's degree of diversification is related to its growth rate of production in its principal product field (textile products in the textile industry and automobiles in motor vehicles manufacturing). In other words, the higher the growth rate of production of an industry's principal products, the lower its degree of diversification and less drastic is its change (motor vehicles, for example). On the other hand, the more saturated an industry's production of its principal products, the higher its degree of diversification and its change (textile, for example). This suggests that diversification may be used as a survival strategy for a declining industry. This is especially true in Japanese industry, where the

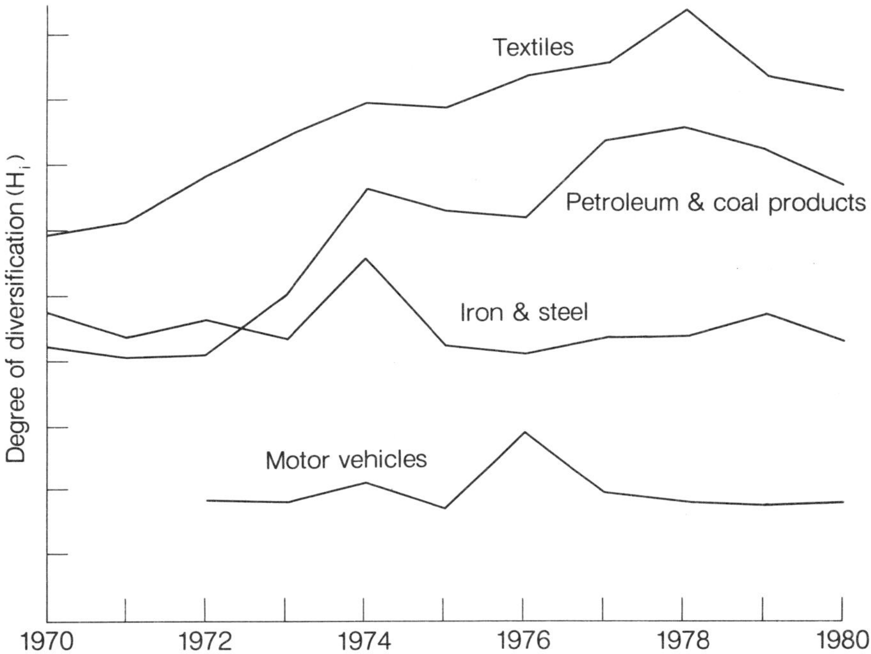

Figure 2.3 Measurement of diversification of major industries

life employment system is built into management practice; managers are motivated to diversify to keep the employment constant.

5.2 R&D investment for diversification

Since an industrial sector's technological diversification is defined as the sector's R&D activity outside its principal product fields, principal product fields must be distinguished from those that are not principal.

In the Statistics Bureau's Survey, however, any given company's R&D expenditures while disaggregated by product field are not disaggregated industry by industry but are assigned entirely to the firm's primary industry. However, many Japanese companies operate in several industries. Therefore, in order to overcome this problem, the twenty-five industrial sectors in the Survey are aggregated into the twenty-one sectors in our analysis. For example, electrical machinery, equipment and supplies manufacturing and communication/electronic equipment manufacturing are aggregated into the one industrial sector of electrical machinery manufacturing. Industrial chemicals manufacturing, oils and paints manufacturing and other chemical product manufacturing are aggregated into the one sector of chemical products manufacturing. On the other hand, transportation equipment manufacturing is disaggregated into motor vehicles manufacturing and other transportation equipment manufacturing, because the major industry of other transportation equipments is the shipbuilding industry, as far as the Japanese statistics are concerned. And transport, communication and public utilities is dismissed in our database, because it is a service industry and covers many different companies from railroad service through electric utility.

Following this reorganization of industrial sectors, all the product fields are classified into the principal product fields of one of the twenty-one industrial sectors, as shown in Table 2.4. This classification is not difficult, because almost all the names of product fields are identical to the names of industrial sectors.

Each sector's R&D expenditure outside its principal product fields (denoted by B) and its relative size in terms of its ratio to total expenditure (denoted by A) is calculated as shown in Table 2.5. The amount of R&D expenditure of all the industries outside their principal product fields was 447 billion yen in 1980 and corresponds to 19 per cent of their total R&D expenditure.

The industries where the expenditure outside their principal product fields exceed those for the principal ones (B/A is above 0.5), are mining, textiles, fabricated metal products and other transportation equipment. The industries in which both types of expenditures are comparable are petroleum/coal, ceramics and nonferrous metals. The industries in which the expenditures outside the principal product fields are minimal compared to those for principal product fields (B/A is below 0.1), are agriculture, drugs/medicines and motor vehicles.

The trends of progress in diversification after 1980 as far as major industries are concerned are shown in Table 2.6. The diversification of the textile industry reached as high as 70 per cent, although its speed has saturated. On the other hand, the diversification of motor vehicles manufacturing remains still quite low, and it will probably not change drastically in a near future. The industry which has changed most in terms of diversification is the iron and steel industry. Its degree of diversification has reached almost 50 per cent, while during 1970s it was only about 20 per cent.

One reason why diversification has been delayed in the iron and steel industry may be attributed to a specific technological feature of this industry: mono-technology culture. The steel industry is a typical mass-production industry, the technology is based on a well-established manufacturing process with few variations available. Therefore, we can assume that the mono-cultural aspects of corporate and technology strategy in this industry prevented diversification efforts compared to

Table 2.5 Measurement of diversification for 1980

Industrial sector	A, total R&D expense (¥ bil.)	B, expense outside principal products (¥ bil.)	B/A
agriculture	2.2	0.10	0.04
mining	8.9	6.25	0.70
construction	53.0	9.55	0.18
food	59.6	19.34	0.32
textiles	17.9	9.64	0.54
pulp and paper	13.1	2.42	0.19
printing and publishing	4.8	1.24	0.26
chemical products	285.3	45.58	0.16
drugs and medicines	162.3	10.87	0.07
petroleum and coal	24.7	11.28	0.46
rubber products	42.9	6.33	0.15
ceramics	62.1	26.18	0.42
iron and steel	188.2	26.23	0.22
nonferrous metals	37.5	15.48	0.41
fabricated metals	38.4	20.36	0.53
ordinary machinery	154.3	37.60	0.24
electrical machinery	666.3	74.55	0.11
motor vehicles	370.0	27.47	0.07
transport equipment	71.1	47.32	0.67
precision equipment	66.4	19.34	0.29
other manufacturing	60.4	30.40	0.50
total	2319.4	447.16	0.19

other declining industries. However, this finding seems to confirm our hypothesis that diversification is used as a survival strategy for a declining industry.

Table 2.6 Percentage of R&D expenditure into product fields outside principal ones to total R&D expense, 1980–6

Industrial sector	1980	1983	1984	1985	1986
textiles	**70%**	**70%**	**72%**	**72%**	**71%**
iron & steel	**22**	**27**	**29**	**34**	**49**
ordinary machinery	24	29	29	33	35
electrical machinery	11	10	10	10	9
motor vehicles	7	7	7	7	7

Source: Statistical Bureau, 'Survey of research and development'.

6. Sectoral patterns of diversification

6.1 Typology of diversification

Industries can diversify their R&D activities in a number of directions. Our purpose is to identify these directions in relation to industrial structure. Therefore, it is important to see how the industries are interrelated. One interrelation that can be identified empirically is that of input to output.

If an industry diversifies its R&D activities into product fields that are either inputs or outputs of that industry, the direction of diversification is called *vertical*. If an industry does not diversify through the path of input-output relations, the direction is called *horizontal*.

There are two types of vertical diversification: *upstream* and *downstream*. If the industry's R&D activities are diversified into product fields that are inputs, the direction of diversification is *upstream*. If they are diversified into the outputs, the direction is *downstream*.

6.2 Measurement of direction

The measurement problem can be formulated as follows: to what extent does the direction of R&D investment, which is given by the distribution of R&D expense into product fields, follow upstream or downstream direction (as given by the input-output transaction table)?

Given

E_{ij} = *i-th* industry's expense into *j-th* industry's principal product fields, for $i,j = 1,2,\cdots,n$.

Let $E_{ii} = 0$, because we are interested in diversification, that is, expenses outside principal product fields.

Then, the *i-th* industry's direction of R&D investment can be described by the vector

$$\mathbf{e}_i = [e_{i1}, e_{i2}, \cdots, e_{in}], \text{ where } e_{ij} = E_{ij}/\Sigma_j E_{ij}.$$

Given

T_{ij} = the amount of transaction from *i-th* industry to *j-th* industry.

Let

$T_{ii} = 0$,

then, the *i-th* industry's upstream direction is represented by the vector:

$$u_i = [u_{1i}, u_{2i}, \cdots, u_{ni}], \text{ where } u_{ij} = T_{ij}/\Sigma_i T_{ij}.$$

Therefore, U_i, the degree of upstream diversification of the *i-th* industry can be measured by

$U_i = \mathbf{u}_i \cdot \mathbf{e}_i$, the inner product of the two vectors.

In other words, the higher the U_i value is, the closer is the direction of *i-th* industry's R&D investment to the *i-th* industry's upstream direction.

Similarly, the *i-th* industry's downstream direction is represented by the vector

$$\mathbf{d}_i = [d_{i1}, d_{i2}, \cdots, d_{in}], \text{ where } d_{ij} = T_{ij}/\Sigma_j T_{ij};$$

then, D_i, the degree of downstream diversification of *i-th* industry can be measured by

$$D_i = \mathbf{d}_i \cdot \mathbf{e}_i.$$

The calculated values of the degree of upstream diversification and of downstream diversification are shown in Table 2.7.

Table 2.7 Measurement of direction of diversification, for 1980

Industrial sector	Degree of diversification	Direction of diversification	
	(B/A)	Upstream (U_i)	Downstream (D_i)
agriculture	0.04	0.396	0.811
mining	0.70	0.032	0.033
construction	0.18	0.042	0.059
food	0.32	0.055	0.099
textiles	0.54	0.328	0.026
pulp and paper	0.19	0.203	0.137
printing and publishing	0.26	0.047	0.057
chemical products	0.16	0.026	0.058
drugs and medicines	0.07	0.235	0.205
petroleum and coal	0.46	0.035	0.154
rubber products	0.15	0.136	0.424
ceramics	0.42	0.058	0.073
iron and steel	0.22	0.032	0.179
nonferrous metals	0.41	0.031	0.146
fabricated metals	0.53	0.024	0.079
ordinary machinery	0.24	0.119	0.111
electrical machinery	0.11	0.063	0.243
motor vehicles	0.07	0.083	0.118
transport equipment	0.67	0.274	0.010
precision equipment	0.29	0.074	0.247
other manufacturing	0.50	0.402	0.075

Among the fairly diversified industries (the ratio of B/A is higher than the average 0.19), the industries with the most strongly developed path of upstream technological diversification are textiles, pulp and paper products, and other transportation equipment. The industries with the strongest path of downstream diversification are petroleum and coal products, iron and steel, nonferrous metals, and precision equipment. However, there are some industries without any strong path of vertical diversification – neither upstream nor downstream. Because they are diversified, we assume they follow the path of horizontal diversification. These are mining, printing and publishing, ceramics, and fabricated metal products.

It is to be noted that the degree of upstream and downstream diversification measured through this analysis is only concerned with the technological diversification, because our database is the allocation of R&D expenses into product fields. Technological diversification does not always lead to product diversification. For instance, the upstream industry such as iron and steel manufacturing can make a technological diversification toward the further upstream direction, by means of being engaged as an engineering company. In order to engage in engineering business, the industry has to diversify its R&D activities into those product fields that are its industrial inputs. Indeed, the active membership of ENAA (Engineering Advancement Association of Japan) except engineering companies and construction companies is composed of the major iron and steel companies and ship building companies.

6.3 Identification of sectoral patterns

There are some industries with the strong path of both types of diversification: upstream and downstream, such as drugs and medicines and rubber products. In other words, although two U_i's or two D_i's can be compared to each other, U_i cannot be directly compared to D_i. In order to make them comparable, both U_i and D_i have to be normalized.

The normalized value of U and D can be calculated by

$$N_u = (U - M_u)/S_u, \text{ and } N_d = (D - M_d)/S_d,$$

where M_u and M_d are the average of U and D, respectively, and where S_u and S_d are the standard deviations of U and D, respectively.

By plotting of (N_u, N_d) in the plane, we can classify sectoral patterns of technological diversification into four categories, as shown in Figure 2.4.

Quadrant 2 includes industry that has strong downstream diversification (its D_i is above the average) but weak upstream diversification (its U_i is below the average). Therefore, the sectoral pattern in this quadrant is identified as downstream diversification. By the same token, quadrant 4 includes industry that has strong upstream but weak downstream diversification. Therefore, the pattern in this quadrant is identified as upstream diversification.

Quadrant 3 includes industry that has both weak upstream and downstream diversification and hence is supposed to have strong horizontal diversification. Therefore, industry in this quadrant is identified as having horizontal diversification. Finally, quadrant 1 includes industry that has both strong upstream and downstream diversification. The sectoral pattern can be identified as vertical

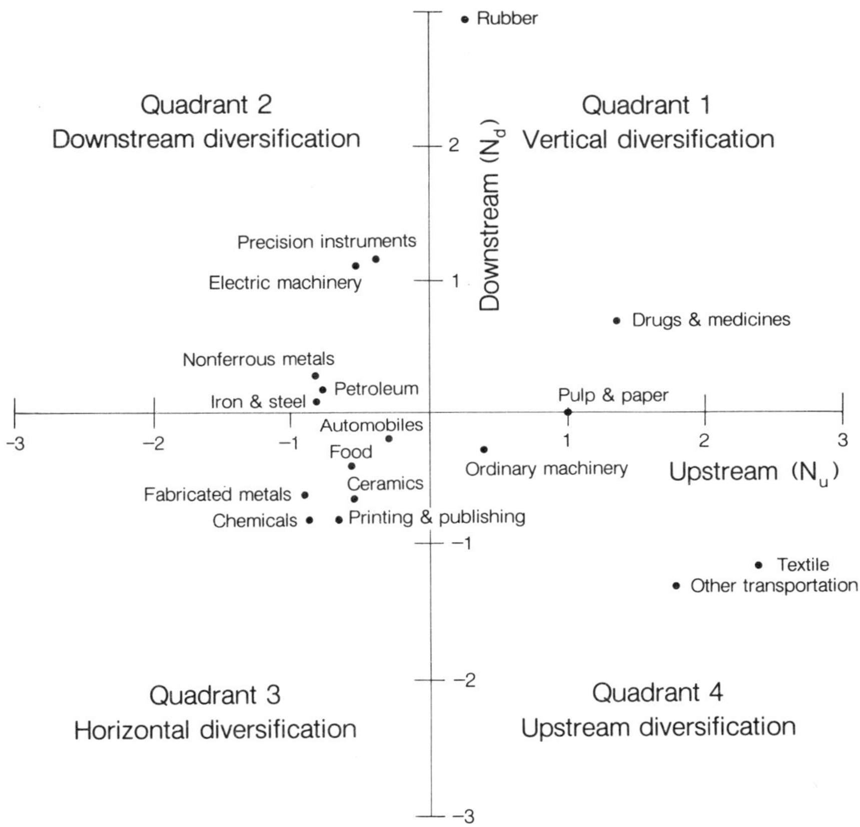

Figure 2.4 Identification of sectoral patterns

diversification. The normalization and classification were made for manufacturing industries.

The industries that can clearly be classified as upstream diversification are textiles and other transportation equipments. They are more or less mature industries, in the sense that they are being caught up by newly industrialized economies. Therefore, they are diversifying toward the upstream direction in order to maintain international competitiveness.

Those clearly classified as showing downstream diversification are electrical machinery and precision equipment. These were the first and second in terms of the growth rate in the period after the oil crisis. Moreover, the mechatronics revolution – the combination of mechanical technology and electronics technology – which occurred around 1975 in

Japan, widened applications for these technologies and hence induced them to diversify into downstream sectors. [13][14] We might generalize that a drastic growth of industry can be only fostered by downstream diversification.

Those that can be clearly classified as showing horizontal diversification include chemical products and fabricated metal products. Since they are typical material industries, their diversification is rather free from industrial structural relations. Drugs and medicines, where the introduction of biotechnology is supposed to be promising, is identified as showing vertical diversification – both upstream and downstream. Those whose direction of diversification cannot be clearly identified are motor vehicles and ordinary machinery.

7. Policies for diversification

7.1 Japanese management and diversification

It is often said that chemical products manufacturing and electrical machinery manufacturing are similar because both are examples of so-called science-based industry. [15] Therefore, it is argued that a country whose ratio of basic research in its total R&D budget is high and that of public science in total R&D is also high has advantage in these two industries. The Japanese ratios are the lowest among the advanced countries.

However, the investments are different; Japan has a strong electronics industry but not a strong chemical industry. The industries are also different in technological diversification, that is, electrical machinery manufacturing is classified as downstream diversification, and chemical products manufacturing is classified as horizontal diversification.

Although both industries are science-based, basic research in the electronics industry has a clear-cut direction, whereas basic research in the chemical industry is undirected, requiring broad-based support in basic research. Therefore, in the electronics industry, management can direct an innovation cycle from basic research to commercialization, using a team approach. Management of basic research is difficult in the chemical industry because managers have to rely on the contributions from a few very talented individuals.

The pharmaceutical industry, which used to be quite similar to the chemical industry, has changed because of the emergence of biotechnology. The direction of diversification in the former has become more clear, that is, a vertical diversification scheme; the chemical industry still follows horizontal diversification. This may be one reason why it is suggested that Japan may soon become the leading competitor of the United States in pharmaceuticals. [16]

7.2 Trade policy

In the 1960s, Japanese governmental policies had been focused on export promotion; in the 1970s, removing trade barriers was central. [17][18]

Therefore, our database available every year since 1970 makes it possible to analyze industry response to the trade liberalization policies as far as the diversification is concerned.

To test the assumption that an industry's loss in international competitiveness leads to upstream diversification, [19][20] industries whose export growth rate is below average (the annual growth rate of Japanese industry was 15.58 per cent in the 1970s) were selected; then industries whose intrinsic nature does not fit exporting (those whose export ratio to domestic production is less than 3 per cent) were excluded. Eight industries were selected as the sample for the statistical investigation.

We have tried to explain their degree of upstream diversification (U_i), that was already calculated in Table 2.7, by means of their export growth rate (E_i), as shown in Figure 2.5. The regression equation was as follows;

$$U = -0.038 \cdot E + 0.58, \; R = -0.917.$$
$$(8.16)$$

Thus, upstream diversification is highly related to export growth, that is, the lower an industry's export growth rate, the larger is the industry's

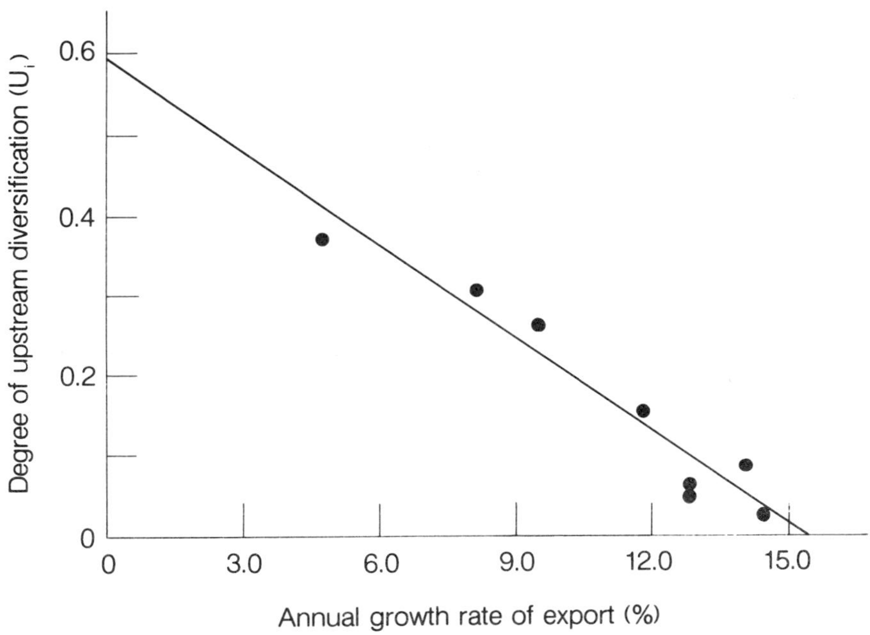

Figure 2.5 Relation between export growth and upstream diversification

degree of upstream diversification. This assumption seems to be true statistically.

Although there are several arguments, this relation may be one reason why the loss in market does not lead directly to a protectionist movement in Japan; with upstream diversification employment can be maintained at individual firms. And this upstream diversification can be a strategy for a declining industry to use to keep its competitive edge against newly industrialized economies, because it could compensate the loss in final product market by the export of products that are its industrial inputs and by plant engineering.

However, we are not claiming that upstream diversification is a new idea or a new Japanese strategy, as it had been already mentioned in several papers such as the Vernon's product cycle work. [21] Nonetheless, this strategy is ascertained statistically by means of our empirical analysis of diversification.

7.3 Industrial policy

In the past, the science and technology policy of MITI was formulated and implemented, not overtly but implicitly, within the framework of industrial policy. [22] Therefore, we can suppose that the industry's technological diversification behavior can be induced by industrial policy.

As we mentioned above, the mechatronic revolution stimulated downstream technological diversification in some industries. Therefore, let us look into the development of industrial policy that brought about the mechatronics revolution of the mid-1970s.

Before 1971, the two laws to promote electronics and mechanical industries (The Law on Temporary Measures for the Development of Machinery Industry enacted in 1956 and The Law on Temporary Measures on the Development of Electronics Industry enacted in 1957), had been enforced independently. The main content of these laws was the low-interest loan given by semi-governmental financial institutions such as Japan Development Bank to the companies for their technology development projects, particularly for their commercialization projects of new technologies.

However, in 1971, these two laws were made into the one law, The Law on Temporary Measures for the Development of Specified Machinery and Electronics Industries. The mechatronics revolution was clearly envisioned in this law, because expressions such as 'consolidation of machinery and electronics into one' or 'systematization of them' were used by MITI for explaining the objectives of the law in the Diet. [23]

In 1978, this law was enhanced into the Law on Extraordinary Measures for the Development of Specified Machinery and Information Industries, which expired in 1985. Although there are some doubts about how effective the influences of these laws were on industry behavior, the merger of the two laws had at least an indicative meaning to the industries involved, in the sense that they showed clearly where major innovations were to occur in the 1970s

7.4 R&D policy

In Japan, a research association scheme is heavily used as the vehicle for the implementation of the governmental R&D policies. [24] This is a joint research arrangement of limited duration in which competing firms share researchers and costs and the government provides funds and tax benefits. This scheme was transferred from the Britain to Japan and reformulated as Engineering Research Association (ERA, Kenkyu-Kumiai) in 1966. On the other hand, MITI launched the National R&D Program (popularly known as the large scale project) in 1966. This was the first attempt by MITI to finance 100 per cent of the cost of R&D which is carried out by private enterprises. Until 1970, the ERA scheme was not used as a vehicle for implementing the large scale project. However, it was realized that the large scale project scheme is compatible with ERA scheme and that they compensate each other in some kinds of research. Thus, it has become normal practice to use ERA for the implementation of MITI's R&D programs. [24]

In the government subsidy program, implemented through the research associations, there seem to be various types of built-in mechanisms to accelerate technological diversification of firms that participate in an association. One mechanism is concerned with venturing. In Japan, unlike the United States, [1][2] ventures are in-house as a rule; not too long ago, I proposed that ERA's were being used by big Japanese firms as springboards into in-house ventures.

It is standard practice for a participating firm to set up an in-house project team that has roughly the same number of members as the research team that the firm sends to take part in the joint research. The project team supports the colleagues on assignment, and it also assimilates the data generated by the ERA. As I see it, the project team can be considered as a sort of in-house venture unit. When the ERA eventually disbands and the employees on assignment return to the company and add their weight to the project team, the team serves, in effect, as the headquarters of a venture capital business. [26] Indeed, the choice of an ERA research subject is often related to the marginal product fields of participating firms, so that it causes less conflict of interest among them. [27]

As far as the direction of diversification is concerned, we can derive some lessons from the experiences of the VLSI project. [25] In this ERA, in which all the participating firms were chipmakers, development centered on how to manufacture the equipment used to manufacture chips and on research into the crystallization process of the silicon, not on how to manufacture the chips themselves. The ERA may be interpreted as a forum for users of manufacturing equipment and silicon crystals to determine together what sort of manufacturing devices and materials should be developed. In fact, many of the development projects were subcontracted to supplier companies in various industries, none of which were members of the association. Thus this joint research led to the creation of several equipment manufacturers and crystal suppliers and the entrance of existing firms into those product fields, which hence helped to build an engineering infrastructure of VLSI manufacturing in Japan.

References

[1] E. Roberts, 'New ventures for corporate growth', *Harvard Business Review*, vol. 58, no. 4, pp. 134–142, 1980.

[2] E. von Hippel, 'Successful and failing internal corporate ventures: an empirical analysis', *Industrial Marketing Management*, vol. 6, no. 3, pp. 163–74, 1977.

[3] F. Kodama, 'Preconditions for flexible manufacturing system: technology, industry, and culture', *Technova Mechatronics News*, vol. 1, no. 1, (Technova Inc., Tokyo, 1983), pp. 17–22.

[4] F. Kodama, 'Micro-electronics and the Japanese employment', *Technova Mechatronics News*, vol. 1, no. 3 (Technova Inc., Tokyo, 1983), pp. 15–20.

[5] 'A survey on users of FMS', (in Japanese), *Nikkei Mechanical*, no. 120 (Nikkei-McGraw-Hill Co., Tokyo, 1982), pp. 61–111.

[6] H. Yoshikawa, *Robot and Human* (in Japanese), NHK Books, 1985.

[7] The Job-Employment Research Institute, *A Survey of Impacts of Micro-electronics on Employment* (in Japanese) (Koyou-Shyokugyou-Sougou-Kenkyushyo, Tokyo, August 1982).

[8] F. Kodama, 'Technological diversification of Japanese industry', *Science*, vol. 233, pp. 291–6, 18 July, 1986.

[9] K. Pavitt, 'Characteristics of innovative activities in British industry', *Omega* vol. 11, no. 2, p. 113, 1983.

[10] F. Scherer, 'Inter-industry technology flows in the United States', *Research Policy*, vol. 11, no. 4, p. 227, 1982.

[11] Statistical Bureau, Prime Minister's Office of Japan, *Report on the Survey of Research and Development* (Nihon-Toukei-Kyoukai, Tokyo, 1970–80).

[12] S. Griliches, 'R&D and productivity: measurement issues and econometric results', *Science*, vol. 237, pp. 31–5, 1987.

[13] F. Kodama, 'Japanese innovation in mechatronics technology: a study of technological fusion', *Science and Public Policy*, vol. 13, no. 1, pp. 44–51, 1986.

[14] C. Smith, *Far Eastern Economic Review*, vol. 127, p. 50, 1985.

[15] K. Pavitt, 'Sectoral patterns for technical change: towards a taxonomy and a theory', *Research Policy*, vol. 13, no. 6, p. 343, 1984.

[16] M. Dibner, 'Biotechnology in pharmaceuticals: the Japanese challenge', *Science*, vol. 229, p. 1230, 1985.

[17] K. Oshima, 'Technological innovation and industrial research in Japan', *Research Policy*, vol. 13, pp. 285–301, 1984.

[18] M. Peck, S. Tamura, in *Asia's New Giant*, H. Patrick, H. Rosovsky (eds.) The Brookings Institution, 1976, pp. 525–85.

[19] O. Grandstrand, *Technology, Management and Markets*, p. 50, Frances Pinter, 1982.

[20] C. Oskarsson, *Technology diversification: The phenomenon, its causes and effects: A study of Swedish industry*, Chalmers University of Technology, Department of Industrial Management and Economics, Sweden, 1989.

[21] R. Vernon, 'International investment and international trade in the product cycle', *The Quarterly Journal of Economics*, vol. 80, no. 2, pp. 190–207, 1966.
[22] C. Johnson, *MITI and the Japanese miracle*, pp. 30–1, Stanford University Press, 1982.
[23] MITI, *Electronics & Machinery Industry in Seventies* (in Japanese), p. 50, Tsushou-Sangyou-Chousakai, Tokyo, 1971.
[24] H. Fusfeld, C. Haklish, 'Cooperative R&D for competitors', *Harvard Business Review*, vol. 85, no. 6, p. 60, 1985.
[25] K. Oshima and F. Kodama, 'Japanese experiences in collective industrial activity: an analysis of engineering research associations', in *Technical Cooperation and International Competitiveness*, H. Fusfeld and R. Nelson (eds.), Rensselaer Polytechnic Institute, 1988. pp. 93–103.
[26] F. Kodama, 'Policy innovation at MITI', *Japan Echo*, vol. 6, no. 9 (Japan Publications Trading Co., Tokyo, 1984), pp. 66–9.
[27] F. Kodama, in *Transforming Scientific Ideas into Innovations: Science policy in the United States and Japan*, B. Bartocha, S. Okamura (eds.), Japan Society for the Promotion of Science, 1985, pp. 198–204.

٢

Chapter 3
R&D activities: from visible to invisible competitors

1. Introduction

An empirical approach to defining high technology industries has been widely adopted by the various administrative branches of government in various countries. One of the well-known definitions, for example, is that used by the Bureau of Economic Analysis, US Department of Commerce.

The BEA defines high-tech industry as an industrial sector which satisfies one of the following two conditions: (1) the percentage of the sector's R&D expense in its value-added output is higher than 10 percent; or (2) the percentage of the sector's number of S/Es (scientists and engineers) in its total employment is higher than 10 percent. The Japanese Ministry of International Trade and Industry adopted the similar criteria. ([1]-[3])

Based on these criteria, examples of industrial products which are chosen as high-tech products are: drugs, computers, telecommunication equipment, electronic machines and components, scientific instruments, NC machine tool, Aircraft, and so on.

However, an obvious discrepancy can be found between the procedure of selection and the result of selection. Industrial chemical products manufacturing is not identified as a high-tech industry by either the US or the Japanese government, although it obviously satisfies the two conditions described above. There should be a good reason why it is excluded from the final list of high technology. However, the governments concerned failed to give an explicit explanation for this exclusion.

The major problem with such a mechanical method of definition lies in the fact that it fails to give any essential difference between the definition for R&D-intensive industry and that for high-tech industry. [4] The definition seems to tell us that high-tech industry is nothing but highly R&D-intensive industry.

Furthermore, inappropriate policies might be applied to high-tech industry if a distinction is not made between these two industrial concepts. A policy for high-tech industry would be an extension of policies applied to R&D-intensive industry. A recommendation to policy planners might be something like more money should be spent on R&D and more S/Es should be recruited to promote high-tech industry; and the government should support basic research because private companies cannot afford it any more; and so forth.

High-tech industry can be distinguished from R&D-intensive industry. First high-tech innovation has certain distinguishing characteristics, like waves of innovation and the possibility of leading companies losing their market share to innovators from other sectors. Secondly, there are structured difference in the pattern of research and development (R&D) activity between high-tech industries and the other industries such as science-based and conventional industries.

The typical high-tech pattern is observed in the electronics industry, where several major innovations occur one after another or simultaneously, like waves reaching the shore. In the last forty years, the electronics industry experienced the series of innovations in which the vacuum tube was followed first by the transistor and then by the integrated-circuit and then by the VLSI. [5] With this kind of situation, corporate decisions on investment are not made on the conventional basis of the rate of return; they are made according to the principle of surf-riding. Companies have no choice but to invest to ride the wave of innovation, because if they do not they are left behind by competitors; investment must continue in order just to stay in the market. [6]

Basic high-tech problems are occasionally solved by ideas or technologies born in totally unrelated fields. Japan's first optical fiber, developed by a glass-maker, had problems with mechanical fragility and transmission loss: areas in which glass-making technology could not provide solutions. As a result, other industries stepped in. Therefore, high-tech companies have to monitor not only direct competitors in their own sector but also firms in other industries. In effect, this means that high-tech companies must engage in R&D competition with 'invisible enemies'.

In this chapter, first, two examples are described in order to illustrate structural features of the high-tech industry. Second, an analytical framework for R&D behavior is developed. Then, based on our empirical studies, it is shown that R&D behavior of high-tech industry is structurally different from those observed in conventional or science-based industries. Finally, the implications for corporate and government policy are derived. [7][8]

2. Structural characteristics of high-tech industry

A typical structural characteristics of the research and development activities in high technology industry is vividly demonstrated by the microelectronics industry and by Japanese development in the field of optoelectronics.

One of the characteristics is most obvious in the recent development of DRAM. In the innovation race of MOS(metal oxide semiconductors)-type DRAM, a new product was introduced into market almost every three years. Furthermore, a new product was introduced before an old product finished the learning process. And whenever a new product appeared, it totally replaced old products within six years. Therefore, we observe the phenomenon of 'systematic replacement of products by innovation cycles'.

In the field of microelectronics, waves came to the shore every three years, but at least the competitor was a company within the microelectronics industrial sector. However, some competitor may show up with a result based on a quite different scientific principle that performs much more effectively than existing products. In many cases the competitor is now a company belonging to a different industrial sector. A closer look at the process of technology development in fiber optics in Japan shows that solutions to basic technical problems have sometimes been found in areas outside the conventional industry.

Given this, we can imagine an innovation model under which leaders in technological advances replace their predecessors regardless of industry, and revolutionary innovation coincides with a change of leadership.

2.1 Development of DRAM

As described above, in the innovation race of MOS-type DRAM, a new product was introduced almost every three years, i.e., 1 Kbits memory chip in 1972, 4 Kbits in 1975, 16 Kbits in 1978, 64 Kbits in 1981 and 256 Kbits in 1983.

As is shown in Figure 3.1, the price of the product decreases in time, and the extent of production increases. However, whenever a new product is introduced into the market, production of the preceding product decreases drastically, although its price still continues to

Table 3.1 Changes in the world top ten chip-makers

	1972 1k DRAM	1975 4k DRAM	1978 16k DRAM	1981 64k DRAM	1984 256k DRAM	1987 1M DRAM
1	TI	TI	TI	Motorola	**Hitachi**	**Toshiba**
2	Motorola	Fairchild	Motolora	TI	**NEC**	**Hitachi**
3	Fairchild	National	National	**NEC**	**Fujitsu**	**Mitsubishi**
4	RCA	Intel	Intel	**Hitachi**	**Toshiba**	**NEC**
5	G.E.	Motorola	**NEC**	National	ATT Tech	**Oki**
6	National	Rockwell	Fairchild	**Toshiba**	**Mitsubishi**	**Fujitsu**
7	GI	GI	**Hitachi**	Intel	**Oki**	TI
8	Corning	RCA	Signetics	Philips	TCMC	**Matsushita**
9	WH	Signetics	Mostek	**Fujitsu**	TI	-
10	AMD	AMD	**Toshiba**	Fairchild	Intel	-

Sources: 1972–81 D. Okimoto, [10] 1984–7 Dataquest.

Note: The Japanese manufacturer is shown in bold.

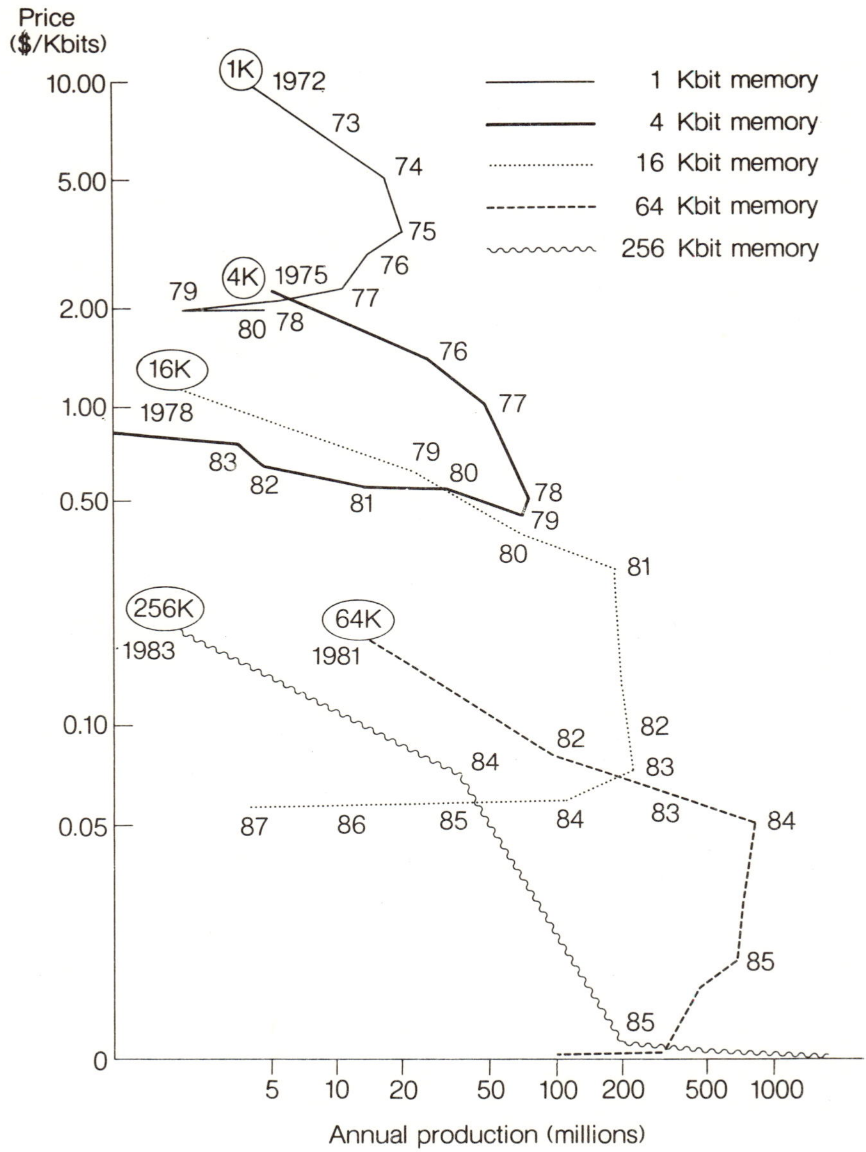

Figure 3.1 The relationship between price and production in MOS-type DRAM (*Source* Dataquest)

decrease. Therefore, manufacturers cannot survive selling only the old products, they must invest in R&D to be able to produce each new generation of DRAM or they are out-of-business

A glimpse at the past two decades' evolution of DRAM as shown in Figure 3.1, demonstrates just how remorseless innovation at the leading edge of technology has been. [9] It shows, too, how destructive it can be to those who fail to catch the wave, as depicted in Table 3.1.

The first commercial memory chip could store 1,000 bits of computer data. When introduced in 1972, the so-called one kilobit (1k) DRAM cost around $10. Initially, around 5 million chips were manufactured annually. The only firms that knew how to make the 1k DRAM were US manufacturers. Between them, they had raised production to more than 20 million chips annually by 1975. Doing so had allowed them to ride the learning curve down to a price of $3 a kilobit.

No sooner had the ten US semiconductor firms started earning profits from the 1k device than its replacement, the 4k chip, hit the market. As shown in Table 3.1, three manufacturers (General Electric, Corning and Westinghouse) missed the wave and were effectively forced out of the business. But three new makers (Intel, Rockwell and Philips) plunged into the game. At its peak in 1978, the 4k device was being produced at a rate of 100 million units a year and priced at 50 cents a kilobit. However, by 1978 the 16k DRAM had become the standard memory chip. This resulted in the pushing up of the cost of 4k in terms of $/Kbits, an unusual phenomena in the theory of learning curve, as can be observed in Figure 3.1.

Three years after the introduction of 16K chip, some 200 million chips were being produced annually for about ten cents a kilobit. Again, three American suppliers (Rockwell, General Instruments and American Micro Devices) failed to catch the wave. Suddenly, there were four Japanese suppliers (NEC, Hitachi, Toshiba and Fujitsu) among the world's top ten memory chip-makers, as shown in Table 3.1.

The Japanese then took the stakes. The 64k device was launched in 1981, a year earlier than expected. At peak production, in 1984, more than 800 million chips a year were produced. Meanwhile, two more Japanese manufacturers had entered into the game. The next wave carrying the 256k chip arrived in 1983, again a year earlier than expected. This time practically all the surfers were Japanese.

Although we can observe that some companies suddenly appeared and disappeared in the race, the underlying trend behind the last two decades' evolution is that the world leadership has shifted from the United States to Japan. Therefore, we can argue that the shift occurred because of the difference in technological and management approach between the two countries, rather than the difference which is specific to individual firms. In other words, we can attribute the dynamics observed in the innovation race to the approach taken by a specific industrial sector in a specific country, rather than to that taken by a specific company.

2.2 The Japanese development of optical fibers

Observation of progress made in the Japanese technological development of optical fibers as well as the transitional passages from research to the production stages reveal that the leaders of innovation changed during technological evolution; that is from glass manufacturers to device manufacturers and from communications enterprises to cable manufacturers, while involving joint research across the boundaries of these industries. Even more noteworthy is that the most dramatic innovations occurred when the leaders emerged from a new sector. [11]

Fiber optics developed by the glass industry lacked mechanical strength and caused transmission loss. The problem of mechanical fragility, for instance, was solved by a coating technology developed by cable manufacturers. The problem of transmission loss, meanwhile, was resolved by NTT (Nippon Telegraph and Telephone Corporation), a major user of optical fibers, which discovered that transmission loss can be greatly reduced by using a wave length much longer than the conventionally used wave length.

In 1969, the joint staff of NSG (Nippon Sheet Glass) and NEC (Nippon Electric Company), in Japan, produced a gallium-doped fiber that had transmission loss of 80 dB/km. In the United States, Corning Glass Works (CGW) produced a TiO_2-doped fiber, that had a transmission loss of 20 dB/km, earning the company worldwide acclaim. NSG of Japan and CGW of the United States were foremost among those companies that made great contributions in the period when the potential of optical fibers was being researched. It should be noted here that the transmission loss of 20 dB/km is comparable to that of coaxial cables.

In the following years, an optical fiber with a transmission loss equivalent to that of guided millimeter wave transmission lines was developed. The development stage now aimed at verifying the usefulness of optical fibers in applications where coaxial cable and guided millimeter wave transmission systems had been used. The leaders of innovation at this stage were the research laboratories of telephone companies.

However, other research areas remained, such as verifying the mechanical strength of fibers before they were practically used, and their ability to endure operation in actual working environments. Those who played a leading role in this research were cable manufacturers that had previously been producing copper wires for communications media. The problems to be solved before the fibers could be applied were overcome by coating technology. Technology known as Tandem Double Coating was developed by Sumitomo Electric Industries (SEI) in 1974. Thus, the leaders during this verification period were cable manufacturers and telecommunications companies.

At the time of practical application, the problem centered around past records of use and system reliability. However, it was risky to attempt to use the system based on this new technology alone. Therefore, the leading players at this stage were the telephone operating companies who would most benefit by the spread of the system and companies engaged in government-sponsored research and development projects. The first of these was the joint research team of NTT and SEI, established in 1975. Their joint research revealed that reductions in transmission loss could be better attained by using a wavelength of, for example, 1.3 μm, which was longer than the previous wavelength of 0.85 μm achieved with GaAs semiconductor lasers. This discovery extended repeater-to-repeater distance and eliminated the need for high-power lasers, thus making fiber optics competitive.

At the same time, it opened the way toward the development of new semiconductor lasers, such as the InP-type, which is capable of lasing in long wavelengths at normal temperature. In this way, optical communications proved to be better suited for long-distance communication than any of the other systems available. Further, in 1979, NTT confirmed a transmission loss of as low as 0.2 dB/km at 1.55 μm, thus giving sure foundation for practical application.

And in 1978, the HI-OVIS system, sponsored by the Japanese Ministry of International Trade and Industry and promoted by SEI, Fujitsu and Matsushita, was completed. This project was a technology experiment and demonstration project of a community communication system in a two-way conversational mode using optical fiber as a means for communication line. At around the same time, the NEC-SEI group from Japan completed the Disney World project in the United States, which commercialized the optical fiber industry. Thus the initial procurement was created artificially and intentionally by the government-sponsored experimentation projects and the initial market was provided by the entertainment business, which was obviously not the most serious customer in terms of various aspects such as cost, reliability, scale and so forth.

With the public beginning to recognize the capabilities of the fiber-optic system and becoming convinced to some degree of the reliability of the system, demand rose substantially. This stage required technology for mass production of optical fibers. The requirement was met by Bell Telephone Laboratories (BTL) and CGW, both of the United States, who developed MCVD (modified chemical vapor deposition). In Japan, NTT and SEI provided VAD (vapor phase axial deposition). Consequently, the contribution to mass-production of optical fibers was made by telephone company laboratories and cable manufacturers. In other words, the principal players at this stage were companies promoting the fiber-optics business and the research organs of the would-be largest users of the system.

In this way, the optical fiber established its position as the communications media of tomorrow's advanced information society. However, a review of this history has revealed that at each stage the leading innovators came from different sectors while involving joint research across industry boundaries. Note also that each time a change occurred, dramatic improvements in technological development were made.

A cyclic process of research, development, production and distribution is called an innovation cycle. [12] As high-tech R&D projects are being carried out simultaneously in a wide variety of industries, the innovation cycle should be thought of as being multi-layered. Leadership changes among players leading technological advances could be taken to mean that leaders move forward from an innovation cycle in one industry to a new cycle in a different industry, settling technological problems along the way. To put it another way, we can think of a spiral innovation model with three-dimensional cycles for high-tech industries. [11] The essential feature of this innovation model is a one-to-one correspondence between technological approach and industrial sector. In other words, each industry tries to solve the problem using the specific technology which is accumulated in the specific industrial sector.

3. The dynamics of R&D programs

It is widely held by students of innovation that innovation is a dynamic process. [13]-[15] However, there is a lesser degree of recognition about the dynamic nature of research and development (R&D) activity.

A well-established classification scheme for R&D activity is not a dynamic characterization of R&D activity, but a static one. It divides R&D into: basic research, applied research and development research. The Department of Defense (DOD) in the US government, unsatisfied with the conventional classification, developed its own scheme.

For the purpose of effectively managing its R&D activities, the DOD divides R&D into: fundamental research, exploratory development, advanced development, engineering development and operational systems development. [16] In management, considerations of cost become very important, especially because of the skewed cost distribution among various stages of R&D activities. The DOD found that the total cost of R&D is distributed as follows: 5 percent is allocated for fundamental research, 10 per cent for exploratory development, 20 per cent for advanced development, 50 per cent for engineering development and 15 per cent for operational systems development. Although there might exist various classification schemes established for management purposes at various institutions, what is important here seems to be the demarcation between the exploratory phase and the development phase.

Generally speaking, an analysis should be quantitative and proven statistically, in order to be scientific. However, as far as the study of R&D dynamics is concerned, there exists, surprisingly, no such analysis, although there are some conceptual models which are neither quantitative nor statistical. [14][17] One of the major reasons for this is that there is no statistical data which reflect the dynamics of R&D activity.

In order to make an analysis of R&D dynamics, we have to define the unit of R&D activity. John Enos, a pioneer in the study of the time interval between invention and innovation, identifies the date of an invention as 'the earliest conception of the product in substantially its commercial form' and innovation as 'the first commercial application or sale'. [18] He found, for example, that radio was invented by de Forest in 1912 and its innovation firm was Westinghouse in 1920.

However, invention activity has become more organized recently. Therefore, it has become more difficult to identify the specific name of the inventor. Even in a study by Enos, for example, he described company chemists as the inventor of DDT (in 1939) and J. R. Geigy as the innovation firm (in 1942).

Nowadays, invention activity is even more institutionalized, and hence it becomes almost impossible to identify the specific name of an individual or a firm as inventor or innovation firm in a scientifically meaningful way. Although several specific names of firms are mentioned in describing the Japanese development of optical fibers in the preceding section, the fact is that we had better describe the development by saying that those product and process technologies which solved the problems were the cable technology and the telecommunication technology, not the glass-making technology. Therefore, we should think of inventive activity as being carried out by the industrial sector.

Industrial R&D activity begins with an industrial sector's interests in a certain product and ends up with the establishment of a new key technology in this product field. This chain of R&D activities which are focused on a certain product field can be called an R&D program.

4. Database for R&D dynamics

In the preceding section, we defined an R&D program as an industrial sector's R&D activities which are focused on a certain product field. Therefore, the disaggregation of the R&D activities by industrial sector and by product fields is an absolute necessity for the study of R&D program dynamics.

Again, the rich Japanese R&D data collected in the *Survey of Research and Development* by the Statistics Bureau of the Prime Minister's Office, in which a company's intramural R&D expenditure is disaggregated into different product fields, provides us with the data base for this conceptualization of the R&D program. [19]

However, the distinction between the expense within the principal product fields and outside them is very important. Compared to the R&D expenditures within the principal product fields, a sector's R&D expenditure outside its principal product fields can be assumed not to involve pure improvements of existing products for their market differentiation, but rather to include only an R&D program as defined above.

These investments outside the principal product fields can be easily canceled if the prospects are not favorable, because at that stage they are investments outside the main business activity. Such cancellation, however, is not so easy in the case of investment within the principal product fields. On this basis, a sector's R&D expenditure in a certain product field outside its principal product fields can be supposed to correspond to a single R&D program, because the investment is focused

Let

D_t : research expenditure deflator of *t-th* year (the reference year is 1975),

then, the real R&D expenditure for various product fields can be represented by,

$$R_{ijt} = E_{ijt} / D_t.$$

This database is available every year since 1970. Therefore, even if the most recent available data were those of 1982 and the number of product field category is reduced to that of industrial sectors by grouping, we can pool a large amount of data, that is, 5,460(=21·20·13) data points for R&D outside the principal product fields, and 273(=21·13) points for R&D within the principal product fields,

The frequency distribution of all the R_{iit}'s (expense within principal product field) reveals that it appears to be lognormal with two peaks. Considering the central limit theorem, the lognormality implies that the sample point includes various types of R&D programs whose natures are different from each other, i.e. it includes various R&D programs, from those aiming at major breakthrough-type innovations to others aiming at nothing but improvement of existing products for their product differentiation in the existing market. Therefore, we can conclude the distribution of R&D expenses within the principal product fields does not reflect the dynamics of an R&D program.

On the other hand, the distribution of all the R_{ijt}'s ($i \neq j$), that is, the expenses outside principal product fields is shown in Figure 3.2. As can be seen, the distribution for the expenses outside the principal product fields is supposed to be exponential.

This exponential distribution in Figure 3.2 implies that almost all of the industry's R&D expenditures outside its principal product fields are for exploratory search and a few of them are for advanced development. We can assume, therefore, the distribution in Figure 3.2 faithfuly reflects the dynamic process of an R&D program, where a heavy investment is realized only after much exploratory search is conducted for a long time and there is evidence of a good prospect. Furthermore, it indicates the very nature of an R&D program, where almost all of the exploratory research projects turn out to be failures and very few of them can survive for advanced development.

Based on the above described reasoning and observation, we will make an analysis of the distribution in Figure 3.2, i.e. that of the industrial sector's R&D expenses into the product fields outside its principal product fields, in order to understand the dynamics of an R&D program.

5. Mathematical formulation of R&D dynamics

In order to construct a dynamic model, we need a dynamic interpretation of the exponential distribution obtained in Figure 3.2. As time passes, an R&D program progresses from an exploratory phase through an

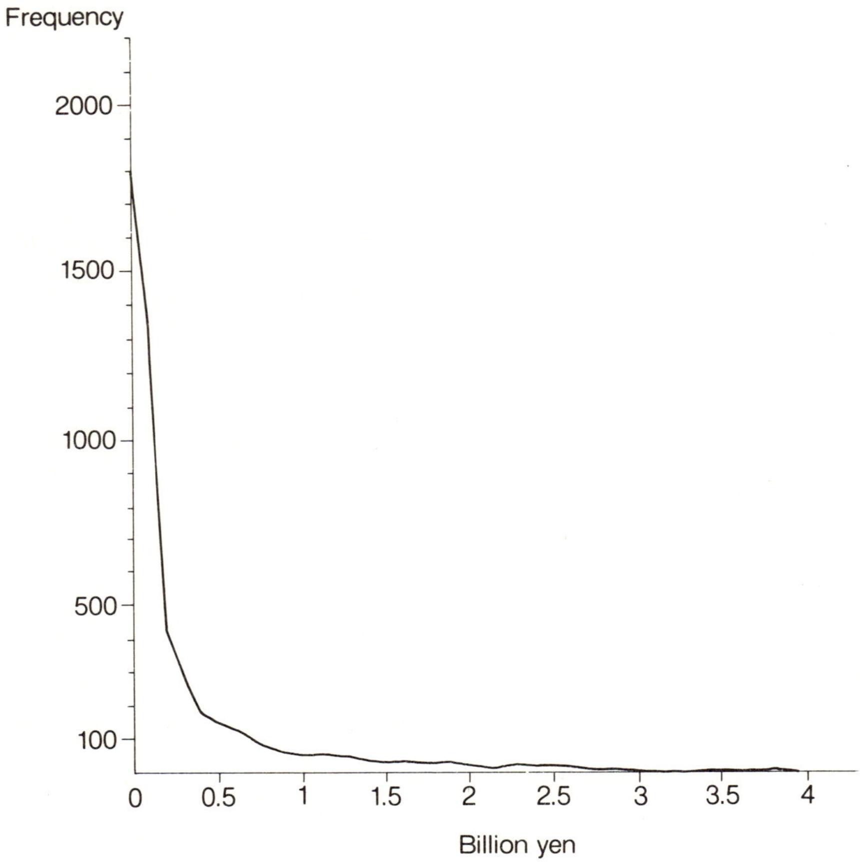

Figure 3.2 Frequency distribution of R&D expenses outside the principal product field (reference year = 1975)

advanced development phase. Thus, its annual investment increases along with the passage of time, as long as its prospect continues to be favorable. However, at anytime when its prospect is found to be unfavorable, it can be canceled, i.e. an increase in its investment is no longer expected.

This suggests that freezing the amount of investment for a research program is equivalent to cancellation of the R&D program, that is, in an R&D program which challenges unexplored fields of knowledge, it is very hard for anyone to conclude that the program has failed, and to decide to eliminate the entire expense all at once. Therefore, research managers will decide only to freeze the research budget.

5.1 Survival model of R&D program

On this basis, we can think of a survival type model of an R&D program's investment as follows:

Let

$R(C)$: the probability that an R&D program can survive until its annual investment reaches the amount of C,

then, the probability that an R&D program is canceled before it reaches the investment level of C, can be represented by,

$$1 - R(C)\cdot$$

Let

$f(C)$: the probability density function of C, i.e. the probability that an R&D program is canceled at the amount of C,

then, $f(C)$ can be represented by,

$$f(C)=d/dC[1 - R(C)] = - R'(C)\cdot \qquad (1)$$

Let

$r(C)$: the canceling rate of an R&D program whose investment level is C,

then, $r(C)$ can be formulated as the conditional probability that an R&D program is canceled, given that it can survive up to the investment level of C, therefore, $r(C)$ can be represented by,

$$r(C)= f(C)/R(C) = - R'(C)/R(C). \qquad (2)$$

If the canceling rate $r(C)$ is constant, i.e. independent of the investment level C, we can let $r(C) = r$, then, the probability that the annual investment level of an R&D program is smaller than C, $R(C)$, can be derived from equation (2) as follows:

$$R(C) = \exp(-r\cdot C)\cdot \qquad (3)$$

Then, the probability density function that the annual investment of an R&D program is the amount of C, $f(C)$, can be derived as follows:

$$f(C) = r\cdot\exp(-r.C). \qquad (4)$$

Therefore, we can derive an exponential distribution of an R&D program's investment, as we observed in Figure 3.2, by assuming that its canceling rate is independent of its investment level.

5.2. Typology of R&D dynamics

It is not always appropriate to assume that an R&D program's canceling rate is independent of its investment level. In other words, we have to conceptualize the canceling rate as the canceling rate function of C.

First, exploratory research can be defined as that where the investment C is smaller, and advanced development as that whose C is larger. Second, the rate at which an exploratory research is canceled can be supposed to be higher, and that of an advanced development to be lower, because a prospect of an R&D program in the advanced development phase is already proven through the exploratory research which precedes it.

Therefore, generally speaking, we can assume that the canceling rate function $r(C)$ is a decreasing function of C, as shown in Figure 3.3.

On this basis, the exponential curve is selected as a function type for the canceling rate function as follows:

$$r(C) = a \cdot \exp(-b \cdot C) \cdot \tag{5}$$

As can be seen in Figure 3.3, the function (5) assumes that the canceling rate approaches zero in advanced development, i.e. there is no

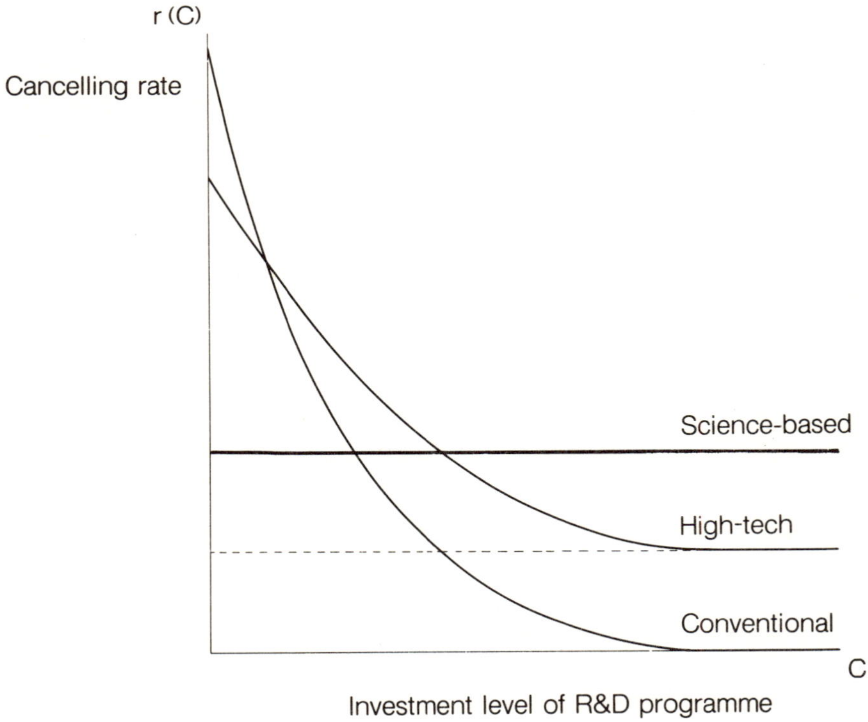

Figure 3.3 Three types of canceling rate function

cancellation once a program enters an advanced development stage. This is a typical pattern observed in conventional industry. Therefore, we call it the conventional pattern.

On the other hand, in the function defined in (3), where $r(C) = r$, the rate of cancellation is independent of the stage of development, i.e. high risks are involved in all the stages of development from fundamental research to engineering development. This pattern is observed in industries where scientific findings are directly related to the main business activity such as industrial chemicals manufacturing. Therefore, we call it the science-based pattern.

Since the function type of (5) assumes that there is no more cancellation once a program enters an advanced stage, there should be several industries where this is not true. Therefore, one more function type is chosen for such industries as follows:

$$r(C) = a \cdot \exp(-b \cdot C) + g, \tag{6}$$

where the canceling rate approaches the non-zero value g, as shown in Figure 3.3, so the possibility of being canceled exists even in the case where a program enters a very advanced development stage.

In this function type, there still remains a certain possibility of being canceled even after an R&D program enters into an advanced development, i.e. development programs have to be canceled if someone else finds another technology which can match the performance level of products based on a different scientific principle. As described before, examples can be found in the electronics industry and in the field of optoelectronics. Therefore, we call it the high-tech pattern.

6. Statistical tests of dynamic models

6.1 Method of estimation

For the canceling rate function, we choose the following three types of function:

type (I) $r(C) = r$ (science-based pattern),
type (II) $r(C) = a \cdot \exp(-b \cdot C)$ (conventional pattern),
type (III) $r(C) = a \cdot \exp(-b \cdot C) + g$ (high-tech pattern).

However, the data for each industrial sector are available only in the form of the probability density function of C. Therefore, we have to derive the probability density function for each type of canceling rate function.

On the basis of the equation (2), we can derive the following two equations:

$$R(C) = \exp\left[- \int_{\phi=0}^{\phi=C} r(\phi) \cdot d\phi \right], \tag{7}$$

and

$$f(C) = r(C) \cdot R(C). \qquad (8)$$

Using these two equations which are applicable for all the three types of canceling rate, we can derive the probability density function $f(C)$ as follows:

For the type (I),

$$\int_0^C r(\phi) \cdot d\phi = r \cdot C,$$

therefore,

$$f(C) = r \cdot \exp(-r \cdot C);$$

For the type (II),

$$\int_0^C r(\phi) \cdot d\phi = a \cdot [\, -(1/b) \cdot \exp(-b \cdot C) + (1/b)],$$

hence,

$$f(C) = a \cdot \exp\,[(a/b) \cdot \{\exp(-b \cdot C) - (b/a) \cdot C - 1\}];$$

For the type (III),

$$\int_0^C r(\phi) d\phi = -(a/b) \cdot [\exp(-b \cdot C) - 1] + g \cdot C,$$

hence,

$$f(C) = [a \cdot \exp(-b \cdot C) + g] \cdot \exp\,[(a/b) \cdot \{\exp(-b \cdot C) - 1\} - g \cdot C].$$

For each of twenty-one industrial sectors, R&D expenses into twenty product fields outside its principal product field are available from 1970 through 1982. Therefore, for each industrial sector, we can use 260(=20·13) data points for the curve fitting of $f(C)$. For the curve fitting, a non-linear least square method, Marquard Method, was used.

Table 3.3 Statistical test of science-based pattern

Name of industrial sector	*r*	Coefficient of determination
food mfg.	0.4467 (26.64)	0.8532
textile mill products mfg.	0.3301 (25.71)	0.8991
pulp & paper products mfg.	0.5628 (15.14)	0.9264
printing & publishing	0.5427 (13.00)	0.9057
industrial chemicals mfg.	0.0830 (37.77)	0.8858
oil & paints mfg.	0.4616 (25.77)	0.8684
drugs & medicines mfg.	0.5269 (21.05)	0.8782
other chemical product mfg.	0.3804 (31.61)	0.9293
petroleum & coal product mfg.	0.4669 (19.09)	0.8786
rubber products mfg.	0.4578 (15.02)	0.8966
ceramics	0.2985 (27.99)	0.8953
iron & steel mfg.	0.2871 (27.04)	0.8865
nonferrous metals & products mfg.	0.2706 (23.08)	0.9193
fabricated metal product mfg.	0.2886 (24.06)	0.9213
ordinary machinery mfg.	0.1535 (18.20)	0.7169
electrical machinery mfg.	0.2301 (42.30)	0.6558
communication & electronics mfg.	0.4232 (86.97)	0.8776
motor vehicles mfg.	0.2753 (49.00)	0.9482
other transportation equip. mfg.	0.3989 (80.57)	0.9452
precision equipment. mfg.	0.3969 (21.31)	0.8738
other manufacturing	0.1688 (18.90)	0.8631

Table 3.4 Statistical test of conventional pattern

Name of industrial sector	*a*	*b*	Coefficient of determination
food	0.4439 (32.15)	0.3955 (8.48)	0.9142
textile	0.3306 (26.36)	0.1889 (5.67)	0.9197
pulp & paper	0.5681 (65.08)	0.5737 (16.14)	0.9965
printing/publishing	0.5309 (12.92)	0.3236 (2.45)	0.9204
industrial chemicals	**0.0844** **(33.68)**	**0.0102** **(1.85)**	**0.8864**
oil & paints	0.4513 (26.68)	0.3220 (6.13)	0.8967
drugs & medicines	0.5376 (66.34)	0.6793 (18.01)	0.9889
other chemicals	0.3840 (43.69)	0.2627 (10.37)	0.9683
petroleum & coal	0.4768 (37.51)	0.5200 (10.42)	0.9722
rubber products	0.4507 (15.19)	0.2681 (3.03)	0.9153
ceramics	0.2995 (28.06)	0.1478 (5.39)	0.9111
iron & steel	0.2968 (35.50)	0.2261 (9.91)	0.9451
nonferrous metals	0.2772 (24.36)	0.1372 (4.80)	0.9388
fabricated metal	0.2991 (30.06)	0.1879 (7.25)	0.9571
ordinary machinery	0.1753 (19.36)	0.1399 (6.36)	0.7894
electrical machinery	0.3029 (100.40)	0.6336 (37.09)	0.9344
com. & electronics	0.4460 (273.30)	0.5211 (79.87)	0.9885
motor vehicles	0.2777 (53.76)	0.1454 (11.06)	0.9643
other transportation	0.3998 (112.10)	0.2680 (25.82)	0.9755
precision equipment	0.4153 (52.28)	0.4657 (15.58)	0.9813
other manufacturing	–	–	–

6.2 The results of estimation

The result of the curve fitting of the science-based pattern (type I) is shown in Table 3.3, with the coefficient of determination and T-value in parenthesis under the estimated value of parameter.

In the table, all the T-values are much larger than 2.02, hence, we can reject the hypothesis that $r = 0$ at the significance level of 95 per cent. However, not all the coefficients of determination are high enough, especially, being low in sectors such as ordinary machinery and electrical machinery.

The results of fitting the conventional pattern (type II) are shown in Table 3.3, with T-values in parenthesis under the estimated parameter values. In the sector termed other manufacturing, we cannot complete the estimation process because convergence conditions are not met. However, we can deny the hypothesis that $a = b = 0$ in all the sectors except industrial chemicals. Furthermore, the coefficients are improved in all the sectors, as compared to those obtained in Table 3.3. However, the coefficient of determination is not high enough in ordinary machinery.

The results of fitting the high-tech pattern (type III) are shown in Table 3.5, with T-values in parenthesis under estimated parameter values. The coefficient of determination is further improved in all the sectors and none of them are lower than 0.8. However, except those sectors such as drugs & medicines, ordinary machinery, electrical machinery, communication & electronics and precision machinery, we cannot reject the null hypothesis that $g = 0$. And in only these five sectors, we can deny that $a = b = g = 0$.

7. Identification of sectoral pattern

7.1 Results of identification

The identification problem here means the choice of a best canceling rate function among the three possible candidates, i.e. type (I), (II) and (III). The criteria for the choice are the degree of fit and the level of significance. This can be implemented by choosing the type whose coefficient of determination is the highest as long as the T-values of all the parameters are larger than the values specified by a significance level.

Based on the decision rule described above, a best function type is chosen. Thus the results of identification are obtained as follows:

These industries identified as science-based pattern, type (I), are:

industrial chemicals;
other manufacturing.

These industries identified as high-tech pattern, type (III), are:

drugs & medicines;
ordinary machinery;

Table 3.5 Statistical test of high-tech pattern

Name of industrial sector	*a*	*b*	*g*	Coefficient of determination
food	0.4411 (29.94)	0.4016 (8.15)	0.0029 (0.52)	0.9145
textile	0.3292 (20.79)	0.1906 (5.30)	0.0015 (0.14)	0.9198
pulp & paper	0.5602 (50.68)	0.5963 (14.28)	0.0083 (1.13)	0.9968
printing/publishing	0.5535 (7.40)	0.2971 (1.92)	−0.0229 (0.36)	0.9212
industrial chemicals	0.1112 (0.32)	0.0062 (0.11)	−0.0274 (−0.08)	0.8865
oil & paints	0.4505 (24.42)	0.3233 (5.94)	0.0009 (0.11)	0.8967
drugs & medicines	**0.5303** **(63.24)**	**0.7069** **(17.88)**	**0.0077** **(2.23)**	**0.9900**
other chemicals	0.3819 (37.51)	0.2664 (9.82)	0.0022 (0.40)	0.9683
petroleum & coal	0.4679 (31.06)	0.5484 (9.77)	0.0095 (1.43)	0.9735
rubber products	0.4575 (9.91)	0.2597 (2.55)	−0.0069 (−0.19)	0.9154
ceramics	0.2991 (21.43)	0.1482 (5.08)	0.0004 (0.04)	0.9111
iron & steel	0.2908 (29.40)	0.2390 (9.25)	0.0065 (1.10)	0.9461
nonferrous metals	0.2776 (15.00)	0.1367 (4.15)	−0.0005 (−0.03)	0.9388
fabricated metal	0.2947 (21.73)	0.1950 (6.47)	0.0047 (0.48)	0.9573
ordinary machinery	**0.1463** **(10.48)**	**0.2474** **(4.28)**	**0.0356** **(2.97)**	**0.8013**
electrical machinery	**0.2764** **(100.30)**	**0.8036** **(35.98)**	**0.0294** **(21.19)**	**0.9569**
com. & electronics	**0.4454** **(272.10)**	**0.5236** **(79.87)**	**0.0008** **(3.32)**	**0.9886**
motor vehicles	0.2762 (43.02)	0.1471 (10.71)	0.0016 (0.39)	0.9644
other transportation	0.3996 (108.00)	0.2683 (25.57)	0.0002 (0.22)	0.9755
precision equipment	**0.4045** **(47.81)**	**0.5027** **(16.23)**	**0.0118** **(2.75)**	**0.9841**
other manufacturing	–	–	–	–

electrical machinery;
communication & electronics;
precision machinery.

These industries identified as conventional pattern, type (II), are:

the other fourteen industries.

Although industrial chemicals manufacturing is a typical science-based industry, other manufacturing is obviously not so. However, it is classified as following this pattern, mainly because it includes so many different industries and the estimation process is not completed in the curve fitting of the other two function types.

7.2 Definition problems of high-tech industries

Generally speaking, there are two alternative ways of defining high technology:

1. based on the industry, from the viewpoint of its technology and R&D-intensiveness; or
2. based on the individual product, from the viewpoint of its attributes and of its technological sophistication.

Although the latter approach is desirable, data availability imposes a limitation on its feasibility. Therefore, the former approach is widely adopted by the various governmental branches in various countries.

One of the well-known definitions, for example, is the one used by the Bureau of Economic Analysis (BEA), US Department of Commerce. Based on the three or four digits of the SIC (Standard Industrial Classification Code), the BEA defines high-tech industry as the industrial sector which satisfies one of the following two conditions:

1. the percentage of the sector's R&D expense in its value-added output is larger than 10 percent; or
2. the percentage of the sector's number of S/Es (scientists and engineers) in its total employment is larger than 10 percent.

Based on these criteria, BEA selected the following products: drugs (283), office, computing & accounting machines (357), electrical machinery (36), aircraft & parts (372), and guided missiles & space vehicles (376), where the corresponding SIC code number is shown in the parenthesis. [1] However, in the competitive assessment study, BEA selected the following industries: drugs & medicines, business machines & equipment, computers, electrical & electronic machines & equipment, telecommunication equipment, electronic components, consumer electronics, jet engines, aircraft, and scientific instruments. [2]

In Japan, the MITI uses the SITC (Standard International Trade Classification) and selected the following products: machine tools (736), automatic data processing machines & units thereof (7521, 7522, 7523,

7528), transistors & similar semi-conductor devices (7764); aircraft & associated equipment (792), and video tape recorder. [3]

Therefore, we can conclude that those industrial sectors classified as high-tech pattern in our study do surprisingly coincide with these selected by the various administrative agencies as high-tech industry, except the aircraft industry which is included in the industrial sector category of other transportation equipment manufacturing together with ship building and railroad vehicles, as far as our study is concerned.

However, in these ways of defining high-technology, there is no essential difference between the definition for R&D-intensive industry and that for high-tech industry. [4] The definition seems to tell us that high-tech industry is nothing but highly R&D-intensive industry.

However, there should be a structural difference in R&D activity between high-tech and R&D-intensive industries. Hence, for the definitional problems of high-technology, we should pay more attention to the structure of R&D activities. This is essentially what was described in this chapter. Therefore, we can claim that our approach is more appropriate than those used by various administrative branches, because our distinction is based on a structural difference.

On the basis of our analysis, we can define the conventional industry as that in which an R&D program has no more danger of being canceled after it reaches an advanced development stage, i.e. risk is deleted once a key technology is established. On the contrary, the science-based industry is defined as that in which the danger of an R&D program being canceled remains at the same level in all stages of its development, for instance, the same level of risk is involved regardless of its development stage. In between, the high-tech industry is defined as that in which the danger of being canceled remains at a certain level even after an R&D program is enhanced to an advanced development stage, i.e. a certain risk remains even if a key technology is established.

8. Policy implications

We have shown that R&D behavior of the high-tech industry is structurally different from those observed in the conventional or science-based industry. Therefore, the policy implication which will be derived from our study will be quite different from the mere extension of the existing policies toward R&D-intensive industry. Let us describe some of the tentative lessons drawn directly from our analysis.

With the emergence of 'high technology', various changes are occurring in the whole framework of research investment decision-making in industry. Investment decisions are no longer based on rates of return. It is more like the principle of surf-riding; the waves of innovations come one after another and you have to invest; if you miss you are left behind. The pattern of competition is also changing; the competitor used to be another company within the same industrial sector, but in many cases nowadays the competitor is a company in a different industrial sector.

Cases in the microelectronics industry concern competitors in technological innovation in the same industrial sector, although competition may be characterized as predatory. However, in the future, solutions to basic technical problems will be sometimes found in areas outside the industrial sector. In other words, basic problems to high technologies will be occasionally solved by ideas or technologies born in totally unrelated fields. Thus, technology-based companies in the high-tech era have to monitor not only direct competitors in their own sector but also firms in other industries. In effect, this means that companies must engage in R&D competition with 'invisible enemies'.

As far as the relationship between R&D management and business management is concerned, we can draw a following lesson. In conventional industry, a key technology is more or less given, therefore the problem is how and when capital investment commercializing the key technology is implemented so that its rate of return becomes highest. Hence, we can separate R&D management from business management.

In science-based industry, on the other hand, R&D management is closely linked to its survival in business. However, this is a sort of management of the unmanageable, because we have to manage basic science. [20] Therefore, in this industry, we can also separate R&D management from business management. Here, a concern in management is how to hedge against high risks. All we can do is to build a strong financial base for a company, in order to support as many basic research programs as possible.

I would argue that this strategy is followed by American and European chemical giants, while it is difficult for the Japanese chemical companies to adopt. It was once reported that the R&D expenditure of the Du Pont Company alone was equivalent to the total sum of the R&D expenditures of the top twenty Japanese chemical companies. [21]

In high-tech industry, however, the situation is quite different from that in these two types of industry, i.e. conventional and science-based industry. R&D activity, especially basic research, needs to be organized and managed, and it can be done to some extent, that is, organized and targeted basic research is more important than random support of basic research. Therefore, R&D management cannot be separated from business management and these two types of management should be consolidated.

It is often said that chemical products manufacturing and electrical machinery manufacturing are similar because both are directly dependent upon basic research. Thus it is argued that a country whose ratio of basic research in its total R&D budget is high and that of public science in total R&D is also high has advantage in these two industries. [22] The Japanese ratios are the lowest among the advanced countries.

However, they are different in international competitiveness; Japan has a strong electronics industry but not a strong chemical industry. The industries are also different in R&D dynamics, that is, electrical machinery manufacturing is classified as high-tech pattern, and industrial chemical products manufacturing is classified as science-based pattern.

Although both industries are directly dependent upon basic research, basic research in the electronics industry has a clear-cut direction, whereas basic research in the industrial chemical industry is undirected, requiring broad-based support in basic research. Therefore, in the electronics industry, management can direct an innovation cycle from basic research to commercialization, using a team approach. Management of basic research is difficult in the chemical industry because managers have to rely on the contributions from a few very talented individuals. In high-tech industries, therefore, systematic management of innovation cycles is an absolute necessity.

As far as the necessity of consolidation of R&D management into corporate management in high-tech industry is concerned, there are various ways of implementation. This can be done, for example, by having as president a manager who has technical insight and prior experience of being engaged in R&D activity. In 1982, for example, all the presidents of the Japanese top five IC manufacturers had science or engineering backgrounds, and some of them had, in fact, extensive prior experience of being engaged in R&D activity.

Generally speaking, I would argue that the general framework of the society may be of some relevance to the development of high technology. For example, Japan developed the societal structure in which the engineering profession is highly appreciated and hence the social status of engineer is quite high. In this society, compared to the other developed societies in the world, the university-trained engineers play a more active role in the decision-making process in every field of society, particularly in corporate decision-making. Through the historical necessity to catch up with the West as rapidly as possible, Japan has developed a higher-educational system in which the schools of engineering are the largest and most popular, attracting the best students in the most prestigious universities. [23]

References

[1] US Congress, OTA, *Technology, Innovation, and Regional Economic Development*, 1982.

[2] US Department of Commerce, *An Assessment of US Competitiveness in High-Technology Industries*, 1983.

[3] MITI, *White Paper of International Trade*, p. 130, 1984.

[4] US NSF, National Science Board, *Science Indicators*, 1982.

[5] J.E. Tilton, *International Diffusion of Technology*, The Brookings Institution, 1971.

[6] F. Kodama, 'How research investment decisions are made in Japanese industry', in *The Evaluation of Scientific Research*, D. Evered and S. Harnett (eds), John Wiley & Sons, 1989, pp. 201–14.

[7] F. Kodama and Y. Honda, 'Research and development dynamics of high-tech industry', *The Journal for Science Policy and Research Management* vol. 1,

no. 1 (The Japan Society for Science Policy and Research Management, Tokyo, 1986), pp. 65–74.

[8] F. Kodama, 'Can empirical quantitative study identify changes in techno-economic paradigm?', in *Science, Technology and Industry Review*, no. 7, pp. 101–29, OECD, Paris, July 1990.

[9] V. Nicholas, 'Thinking ahead: a survey of Japanese technology', *The Economist*, vol. 313, no. 7631, 2–8 December 1989.

[10] D. Okimoto, T. Sugano and F. Weinstein (eds), *Competitive edge: The semiconductor industry in the US and Japan*, Stanford University Press, 1984.

[11] F. Kodama et al., 'The innovation spiral: a new look at recent technological advances', presented to *The Second US–Japan Conference on High Technology and the International Environment*, National Academy of Sciences and Engineering Academy of Japan, Kyoto, Japan, 9–11 November 1986

[12] National Research Council, *International Competition in Advanced Technology: Decisions for America*, pp. 28–38, National Academy Press, 1983.

[13] N. Rosenberg, *Perspectives on Technology*, Cambridge University Press, 1976.

[14] W. Abernathy and J. Utterback, 'Patterns of industrial innovation', *Technology Review*, June/July, 1978.

[15] F. Kodama and S. Tamura, 'Technological Innovation' (in Japanese), *Journal of the Japan Society of Instrument and Control*, vol. 21, no. 3, pp. 377–85, 1982.

[16] NASA, *Phased Project Planning Guideline* (NHB 7121.2), 1968.

[17] R. Evenson and Y. Kislev, *Agricultural Research and Productivity*, pp. 140–55, Yale University Press, 1975.

[18] J. Enos, 'Invention and innovation in the petroleum refining industry', in *The Rate and Direction of Inventive Activity*, Princeton University Press, 1962.

[19] Statistics Bureau, Prime Minister's Office of Japan, *Report on the Survey of Research and Development* (Nihon-Toukei-Kyoukai, Tokyo, 1970–82).

[20] F. Kodama, 'Introduction to research evaluation', (in Japanese) *Communications of the Operations Research Society of Japan*, vol. 28, no. 11, pp. 520–57, 1983.

[21] *Asahi Newspaper*, 'Japan becomes a battlefield', 20 May 1985.

[22] K. Pavitt, 'Sectoral patterns of technical change: towards a taxonomy and a theory', *Research Policy*, vol. 13, p. 343, 1984.

[23] F. Kodama and C. Nishigata, 'Structural changes in the Japanese supply/employment systems of engineers: are we losing or gaining?', in *The Changing University and the Education and Employment of Scientists and Engineers*, NATO ASI SERIES, Kluwer Academic Publishers, forthcoming.

Chapter 4
Technology development: from linear to demand articulation process

1. Introduction

In the high-tech era, the key issue of technology policy has become not how to break through technological bottlenecks, but how to put existing technology to the best possible use. Accordingly, a day of reckoning has come for technology policy, which traditionally has emphasized the supply side of technology development. A need has now arisen for a technology policy which works from the demand side.

In developing new policies to meet this need, the most important element is the process in which the need for a specific technology emerges and the R&D effort is targeted towards developing and perfecting it. We call this 'technological demand articulation'. According to Webster's dictionary, the word 'articulate' comes from the Latin *'articulare'*. Among other things, this word has the following two conflicting meanings: (1) to draw up in separate articles: and (2) to put together by joints. Therefore, this word can be interpreted as encompassing the two opposite concepts: analysis and synthesis, or decomposition and integration. In fact, both are necessary in technology development, and the heart of the problem concerning technology development is how to manage those two conflicting tasks.

Therefore, we define demand articulation as a dynamic interaction of technological activities that involve integrating potential demands into a product concept and decomposing this product concept into development agendas for its individual component technologies.

From the technologists' viewpoint, technological demand articulation is the search and selection process among technical options. However, the sample population from which technical options are to be drawn varies over a wide spectrum of sources of innovation. It may be drawn from the shelf of existing technical collections. It may also be drawn from the pool of scientific knowledge, which is advocated by scientists and technologists in terms of a linear progression model of innovation. Between these two extremes, there exists a wide range of technology development.

Good evidence of the importance of demand articulation and its public policy implication can be seen in the rich Japanese experiences of research consortia, in which rival firms participated. The Engineering Research on a certain product and does not include simple improvement of existing products.

Moreover, this expense can be assumed to be the output indicators of an R&D program, because the increase in expense can be expected due to the success in the earlier stages. However, we cannot make this assumption in the case of the expense within the principal product fields.

First of all, we have to distinguish the two types of product fields for each sector: its principal product fields and those which are not principal. We made this distinction among manufacturing industries. In other words, every product field produced by manufacturing industries was classified into the principal product fields of one of the twenty-one manufacturing industrial sectors. The classification table is shown in Table 3.2.

Table 3.2 Classification of principal product fields for R&D dymanics analysis

Name of industrial sector	Principal product fields
food mfg.	food products
textile mill products mfg.	textile products
pulp & paper products mfg.	pulp & paper products
printing & publishing	printing & publishing
industrial chemicals mfg.	chem. fertilizers and organic & inorganic chemical products Chemical fibers
oil & paints mfg.	oil & paints
other chemical product mfg.	other chemical products
drugs & medicines mfg.	drugs & medicines
petroleum & coal products mfg.	petroleum products
rubber product mfg.	rubber products
ceramics	ceramic products
iron & steel mfg.	iron & steel
nonferrous metals & products mfg.	nonferrous metals
fabricated metal product mfg.	fabricated metal products
ordinary machinery mfg.	ordinary machinery
electrical machinery, equip. & supplies mfg.	household elect. appliances; other electric equipment
communication & electronic equip. mfg.	communication & electronic equipment and electric gauges
motor vehicles mfg.	automobiles
other transp. equip. mfg.	ships; aircraft; other transp. equipment
precision equip. mfg.	precision instruments
other manufacturing	other manufacturing products

Then, we can formulate the data availability of R&D expense as follows:

Let

E_{ijt} : *i-th* industry's R&D expense into *j-th* industry's principal product fields in *t-th* year ($i, j = 1,\cdots,N; t = 1,\cdots,T$),

then, E_{iit} represents *i-th* industry's R&D expense into its principal product fields in *t-th* year.

Association Act, enacted in 1961, established the framework for setting up research associations for a limited duration. In order to ascertain its technological rationale for collective research, we can formulate the problem as follows: how to create and guarantee a situation in which a 'cooperative game' can be structured among rival firms. The findings were that collective action through cooperative games creates the functional equivalent of demand articulation for a technology by a single firm.

Specifically, two associations in the field of engineering research in microelectronics and information technology are selected. They are analyzed using the insights provided through the collective version of the concept of demand articulation. In other words, they are analyzed from the perspective of an overall framework which considers both upstream and downstream national industrial technological linkages.

In order to show the potential operationablity of the concept of demand articulation, we will try to analyze the participation behavior of rival firms using a database which contains the experiences of Japanese collective research. Our analysis is focused on the economic rationale for collective research. The database of the IEA (International Energy Agency)-led international collaborative research is used for the analysis of science projects, and the database of NTT-initiated collective research programs among communication equipment manufacturers is used for the analysis of technology projects. Collective behavior is observed in the frequency-distribution of the number of countries or firms that participated in a project.

It is found that a science project and a technology project produce quite contrasting statistical phenomena. These two different phenomena can be explained by mathematical models of independent versus dependent decision-making on participation. The dependent decision-making model of participation in a technology project is built using the scheme of a 'non-cooperative game'. It is also found that a firm's decision is based on its estimation of benefit/cost ratio. Through this model it is shown that collective action by rival firms is economically rational behavior, although the observed phenomenon is counter-intuitive.

By means of both quantitative and descriptive analysis, we can show that demand articulation is a critical concept in terms of both corporate and governmental policy. From this analysis, several policy implications are drawn in terms of management of research projects and of guiding principles for research policies.

2. Survey of relevant literatures

In order to succeed in the process of demand articulation, you must have good marketing, good economic understanding and good technologies.

As far as the market for technology is concerned, many articles in the field of market research cover this area. They point out the importance of understanding market needs, but few of them mention technology development.

Somewhere outside market research articles, we can find some research which mentions technology development. In the SAPPHO (scientific activity predictor from patterns with heuristic origin) project, which was conducted by SPRU (Science Policy Research Unit) in the United Kingdom, it was pointed out that understanding users' needs is the most critical factor for successful innovation. [1] In some areas, Eric von Hippel of MIT argues, users who understand the needs of the market most often develop the technology first. [2] He proposes a customer-active paradigm of technical innovation; however, his analysis is limited to innovation in industrial goods, such as scientific instruments and semiconductor process equipment. In the area of consumer goods, however, it is hard to imagine consumers being the source of innovation. Because analysis in this field focuses on the factors and actors of innovation, it has several management implications but not many policy implications.

The economics of innovation point to the fact that innovation occurs when it pays, thus stressing the cost-effectiveness side of innovation. Since this framework of analysis is static and retrospective, it does not lead to any implications for technology policy which addresses the dynamic and action aspects of innovation.

Some economists, however, pay attention to the intrinsic dynamics of technology development. Nathan Rosenberg concludes that backward linkage has been an enormously important source of technical change in the Western world. [3] He discusses the manner in which the demand for new techniques emerges and is perceived. He argues that the ordinary messages of the marketplace are general and not specific enough to indicate the directions in which technical change should be sought. Therefore, he argues, besides the marketplace there must be forces pointing emphatically in certain directions. Rosenberg offers the concept of technological imperatives as guiding the evolution of certain technologies: bottlenecks in connected processes and obvious weak spots in products form clear targets for improvement.

On the other hand, as Nelson and Winter argue, the directions taken seem 'straighter' than what is suggested by Rosenberg's emphasis on the shifting focus of attention. [4] They term these paths natural trajectories. The interpretation of trajectories within classes of technology is useful for organizing thinking about certain irregularities in the pace and pattern of technical progress. They also point out that within any of these classes of technology, technological advance may follow a particular trajectory. At any given time all R&D may be focused on one class of technologies, for instance, with no attention being paid to other technologies because of lack of knowledge in that area.

Their approach is, of course, an historical analysis; hence, the cases they dealt with are major innovations which took a long time to occur. Their concepts of 'technological imperatives' and 'natural trajectories' are based on long-term and macro-level phenomena of technical change. However, when it comes to the analysis of technology policy, we need a micro-level version of their concept.

From the technology viewpoint, the demand articulation can be interpreted as the search and selection process among technical options. However, the sample population from which technical options are to be drawn varies over a wide spectrum of sources of innovation. It may be drawn from the shelf of existing technical collections, where marketing and economic research are predominent. It may also be drawn from the pool of scientific knowledge, which is advocated by scientists and technologists in terms of a linear progression model of innovation, although some technologists expressed disagreement on this model. [5] In between these two extremes, there exists a wide range of technology development, which might be best described as 'targeted technology development'.

In endeavors of this sort of technology development, Nelson and Winter's metaphor of 'alternatives out there waiting to be found' is somewhat forced. As far as sources of innovation are concerned, however, as described above, some parts may be drawn off the shelf of existing technologies, while some parts may be drawn from the scientific pools or sometimes beyond them. In fact, this range of technology development may cover the majority of technology development.

3. Structure of demand articulation: integrated circuits technology in the defense sector

The demand articulation concept is most visible in US experiences of IC development in the defense sector. It is said that integrated circuit technology was developed through defense research in the United States. However, a detailed investigation into the development process reveals the importance of demand articulation. In other words, a vaguely defined demand for national security did not induce the development of IC technology; rather, it was due to the successful translation of the defense demand into agenda-setting for technology development.

The defense strategy in the decade after World War II and with the intensification of the Cold War was formulated around the concept of 'deterrence', which replaced the concept of sheer 'destructive power' of nuclear bombs which prevailed during World War II.

In the early years after WWII, the essence of US strategy emphasized the sheer destructive power of nuclear weapons, a strategy that found its essence in Secretary of State Dulles's emphasis on massive retaliation. By the mid-to-late 1950s, the inadequacy of such a simple notion of destructive power strategy was increasingly recognized, and was demonstrated by the unwillingness of the United States to use such weapons or their umbrella, in Dulles's phrase, 'at a place and time of our choosing', in relation to the 1956 Hungarian revolution; the ultimately negative assessment about using US nuclear weapons in the face of the French defeat in Vietnam; and, most importantly, the development of Soviet nuclear weapons and missiles.

These factors combined with the RAND-based study with its emphasis on assured second strike forces and loss of overseas bomber bases, the counter-force preferences of the Kennedy Administration and emerging overhead reconnaissance and targeting technology worked to create a general climate of strategic demand. The demands were for precision in delivery of nuclear weapons, highly survivable missiles suitable for a second-strike, and communications, command and control of nuclear forces that was equally survivable. Also, the demands were for technologies that would minimize missile size and maximize the space and weight that could be used for weapons *per se*, i.e. payload. The United States came to opt for lighter rather than heavier missiles, in contrast to the Soviet Union.

In this context, how quickly and how accurately nuclear weapons could be carried to the targets became the critical factors. Thus the development of the carrier of nuclear weapons, i.e. the missiles, became the central issue. Defense R&D was then focused on the development of small and reliable electronic circuitry to control the missile. Through this process, the defense strategy of deterrence was translated into the technological problems of 'miniaturization and higher reliability of electronic circuitry'.

When it became clear that the requirements of miniaturization and greater reliability of electronic circuitry could not be fulfilled by the conventional vacuum tube technology, nor by the transistor, a breakthrough concept of molecular electronics was suggested by the Department of Defense in 1958. [6] In brief, components using this technology would embody: the various electronic functions without specifically fabricating electronic parts including transistors, diodes, capacitors and resistors; and, the material itself would simulate electronic function such as that of oscillators and amplifiers. Responding to this concept, R&D experiments were conducted in the various institutions and the IC technology was generated.

A point worth noting is that Texas Instruments and Fairchild Semiconductor, who made key innovations in IC technology, sought and received no support from the government for their early IC development work. Texas Instruments demonstrated the first IC as we know it today, i.e. electronic components indivisibly embodied within semiconductor materials. Fairchild provided the planar process that revolutionalized the embryo IC then available and moved it from the laboratory to the production line. These breakthroughs, however, were brought about through the in-house R&D efforts of those companies which responded to the articulated demand of the military.

We can learn the following lessons from this experience in the US defense sector. The government articulated and shaped the problem to which the innovative candidate technology should be addressed, and then supported the more promising of the aspirants generated by industry.

The structure of demand articulation is a multi-level hierarchy. The causal chain of demand articulation contains an overall strategy level, a system requirements level and a technological level. In conclusion, such a complex structure is necessary to articulate the demand for a radical innovation such as integrated circuit technology.

4. Process of demand articulation

In the preceding section, we focused our analysis on the structural aspects of demand articulation. Therefore, in this section, we will try to pay more attention to the process of demand articulation.

In the Introduction, we defined demand articulation as a dynamic interaction of technological activities that involve integrating potential demands into a product concept and decomposing this concept into development agendas for its individual component technologies. The best example of a dynamic interaction of the two kinds of technological activity can be found in the development of liquid crystal technology. The process of integrating potential demands into a product concept can be illustrated by the development of Japanese language word processors. And the process of decomposing a product concept into development agendas for component technologies can be demonstrated by the development of the home-use VTR (video tape recorder). [7][8]

In the case of the liquid crystal technology, what turned out to be the correct technological choice was made by the more specialized manufacturers who limited their applications to displays for calculators and watches, which had a specific need for a reduced energy requirement. In other words, they were the only manufacturers who achieved demand articulation *vis-à-vis* liquid crystal technology.

In the case of Japanese language word processors, the first attempt with ideas of direct analogy from English typewriters ended in failure. Therefore, manufacturers were forced to return to fundamental research, starting from the basic principles of linguistics. The result was the birth of a new product concept: the Japanese language word processor, in which a primitive version of AI (Artificial Intelligence) technology is embodied.

In the case of home-use VTR, a long-term commitment to a specific product concept made a consistent demand articulation possible. This, in turn, allowed the successive development of new technologies to meet specific technological needs. In retrospect, these achievements would have been inconceivable under a mere extrapolation of existing technologies.

However, through all the three case studies, we will find that the successful demand articulation required the following prerequisites. *First*, a given industry's capacity for demand articulation depends greatly on the technological level of related industries. All industries must have a high level of technological capability before a high degree of demand articulation can take place. *Second*, demand articulation requires brisk competition between companies, almost to the point of excess. This competition, moreover, must focus on developing specific technologies for specific needs. Technological competition is ultimately competition over how skillfully demands can be articulated. *Finally*, it should be noted that demand articulation requires a long-term view. This means a long-term commitment to providing stable and adequate financial and human resources for research and development, of which demand articulation is a crucial part.

4.1 Development of liquid crystal technology

In the development of liquid crystal technology, demand articulation for more advanced, sophisticated products helped foster innovative new technologies. There were clear commercial applications for liquid crystal displays, and the companies which were most successful in this field were those whose technological goals were most suitable to such applications.

The watch manufacturer, Seiko, whose overriding objective was the development of an electronic quartz watch, had specific needs for technology which would extend the life of its product, reduce energy consumption and allow for a smaller, thinner display. For that reason, from an early stage the company undertook research into displays using little energy, and soon narrowed their sights to liquid crystal displays.

At the same time, The Sharp Corporation was caught up in stiff competition in the development of pocket calculators. Its key to success was creating smaller, thinner products which used little energy. It would be bothersome, for example, to carry around a calculator whose batteries had to be replaced after every ten hours of use. By focusing its technology development efforts on liquid crystal displays and CMOS (complementary metal oxide semiconductors), Sharp was able to extend the life of its calculators and develop a card-size calculator.

Liquid crystal was discovered by Europeans and they led the basic research community in understanding the principles and developing materials. The basic idea of utilizing liquid crystals as a display device was brought about when RCA invented the dynamic scattering mode (DSM) in 1967. RCA demonstrated various prototype products such as a device displaying numerals and letters, a window-glass curtain, still picture display equipment and a display panel for airplane pilots. All the products, however, were premature in the then available technologies and RCA gave up commercializing them.

Those who developed the liquid crystal display technology and made the manufacturing technology possible for mass production were the Japanese watch and calculator manufacturers. Thus success in these product areas demonstrated to the world the effectiveness of liquid crystals and established the basis for stable, long-life liquid crystal technologies. Thereafter, various innovations followed, bringing larger screens, higher precision, quicker response, color displays and ease of display.

When liquid crystals were invented, the standard for display technology was the cathode-ray tube. The flat panel display was nothing more than a dream. Nevertheless, other technological alternatives to liquid crystals existed, such as electroluminescence and plasma display, and manufacturers agonized over technological choices. RCA, which had itself invented the liquid crystal display, chose to stick with the CRT technology, as did most manufacturers of CRT screens.

4.2 Development of Japanese language word processors

In the development of Japanese language word processors, careful investigations into a cultural factor made demand articulation possible. In this development, adequate time had to be devoted to fundamental research and the careful articulation of needs, because the technology is directly related to the culture of a specific country.

The Japanese language uses a mixture of the 50-character *kana* alphabet and over 3,000 Chinese *kanji* characters. The 'keyboard' of a traditional Japanese typewriter – actually a plate printed with rows of tiny characters, and a mechanical arm with a pointer to select them – contains over 3,000 characters which the operator must find and punch in one by one. Not surprisingly, only trained specialists are able to operate them. Thus the diffusion of Japanese typewriters was far behind the American and European machines. And it was feared that this handicap would become the bottleneck in office automation in Japan.

The first attempt at a Japanese language word processor, which also used a plate printed with individual characters, i.e. the tabulating input equipment, ended in failure. Manufacturers were forced to return to fundamental research, starting from the basic principles of linguistics. After ten years of wide-ranging research at Toshiba's laboratories, a new method was finally devised in which *kana* can be converted to *kanji*. The result was the birth of a new product concept: the Japanese language word processor. Working from a keyboard of forty-eight *kana*, the operator can simply key in text as it is pronounced, while a computer programmed with a dictionary of Japanese vocabulary and grammar automatically converts the *kana* letters into *kanji* characters where necessary.

Development was conducted in related technologies such as dot-matrix printers and cathode-ray tubes for display. In the choice of printer technology, it was clearly recognized that the minimum requirement for a Japanese word processor is 24 by 24 dots, not 16 by 16 dots, which was available at that time. In order to realize the 24 by 24 dots printer, research was done on the diameter and arrangement of the pin of print head and on materials. Thus the new wire dot printer was developed. The other reasons for this choice were the requirements for carbon print and difficulty in making a change in official documents, such as those to be submitted to the government.

The resolution capacity of cathode-ray tubes for display was 600–700 lines, not sufficient for the 1,000 lines which are necessary for A4 size paper and 40 character of 24 dots per line. This problem was solved by enhancing the horizontal resolution of the cathode-ray tube.

In seven years, the price of Japanese language word processors was reduced by 98 percent and its size was significantly reduced. Thus the Japanese language word processor is now widely used not only in offices but at home. The market size of the processor has grown to 2 million units annually.

The development history of culture-specific products provides a contrast because their development must be conducted domestically, and thus one could not expect much of the foreign technological reserve. Moreover, once goals had been set, the technology had to be improved until these goals were fully met. This analysis will also serve to verify the importance of demand articulation in product development in areas other than products which are universal in nature and thus can be distributed worldwide.

4.3 Development of the home-use VTR

In the case of video tape recorders, what made demand articulation possible was the manufacturers' long-term commitment to a specific product concept. Ever since the appearance of VTRs for use by the broadcasting industry, Japanese manufacturers were keenly aware of the future for home-use VTRs. Even in choosing which technologies to use in developing industrial-use VTRs, they took into account the eventual development of home-use VTRs. Moreover, their successes on this point were not the result of choices among existing technologies, but of research which returned to basics, such as the fundamental format for VTRs and component technology.

In the video tapes by the broadcasting industry, the video signals for each frame are recorded in several track lines perpendicular to the length of the tape with several magnetic heads rotating perpendicularly. This movement of the magnetic head causes vibration of the video tape as it runs through the machine. Controlling this vibration was a key technological challenge, because it causes variations in the recording and playing time and hence image jitter on the TV screen.

However, for more compact broadcasting-use equipment and for home-use VTRs, a greater imperative was finding a way to use narrower video tape without splitting the picture, which would cause the screen image to fluctuate. Thus, in 1955, Norikazu Sawazaki of Toshiba invented the helical scanning system: an innovative means of recording signals diagonally across a narrower tape with a helically rotating magnetic head so that a field video signal can be put in a single recording track line without jitter. [9]

In the middle 1970s, when the home-use VTR market was beginning to grow, Toshiba was attempting to develop another new system of horizontal recording. Sony, on the other hand, who had made a long-term commitment to home-use VTR, adopted the conventional helical scanning system and succeeded in articulating the market demand with the helical scan technology, and the standardization process began before Toshiba's new system was perfected. Today, the helical scanning system is widely used in almost every VTR, and thus the Vladimir K. Zworykin Award was given to the inventor Dr Sawazaki from the IEEE (Institute of Electrical and Electronics Engineers) in 1986.

For color VTRs, since the density of colors is expressed through the amplitude of the 3.58 MHz sub-carrier of color signals and color tones are expressed through the phase of the carrier, a slip in time can cause a slip in the phases. In industrial-use VTRs which were developed by Ampex, this is corrected through massive electronic circuitry. This was not feasible for home VTRs. Sony, however, responded by reducing the frequency of the carrier to the one-sixth, 600 KHz, a level at which small time slippages would not affect the color.

In the development of the 8mm VTR, the demand was for a miniaturized and lighter machine. And the fundamental technology was a development of the recording medium of higher density. The existing tapes of iron oxide, which had been used in the last thirty years, were reaching physical limits. On the other hand, metal tape technology was not yet realized, although it was known that its energy density was four times that of iron oxide tape. In order to utilize metal tape in the VTR, various technologies had to be developed, such as a high quality head for the metal tape, a smooth and rust-free surface for the metal tape and a low temperature affinity of the tape base.

As to the development of a head, for example, a metal head was used first, then a ferrite head. The magnitude of the magnetic field of the ferrite head was about 650 oersted, a level far below the 1,450 oersted needed for a metal tape. Therefore, a composite head was developed; the metal was used only for the critical part of heads gap and the ferrite was used for the total magnetic circuitry because its reproduction characteristics are much better.

5. Public policy for demand articulation: Japanese experiences in collective research

One of public policy implications of demand articulation can be found in the rich Japanese experiences in research consortia, in which rival firms cooperate.

In the transition of the IC market from defense to the civilian sector, the development of manufacturing technology becomes more important. In this area, the Japanese manufacturers played an important role. In particular, the Research Association for the VLSI development, which was in existence from 1976 through 1979, made a major contribution. The association included all of Japan's five major IC chip manufacturers at the time: NEC, Toshiba, Hitachi, Fujitsu and Mitsubishi Electric. The association made possible demand articulation for manufacturing equipment and materials for chip-making. [10]

5.1 The scheme for collective research

The research association scheme was first introduced by Britain in 1921, and then diffused to such countries as France, Germany, Italy and Norway. This scheme was transferred to Japan and reformulated as the 'Engineering Research Association' (ERA, Kenkyu-Kumiai) in 1961.

Although the birthplace of this system was Britain, the Japanese version is somewhat different. The research associations active in Europe were intended mainly to support and enhance the technological ability of small businesses lagging behind in modernization. The purpose of Japanese cooperative research, by contrast, was to raise the technological expertise of major firms to the top of the world level. [10]

Since the Japanese associations are aimed at reaching the most advanced technology possible, the research objectives are set a generation ahead of the existing technology. Current differences in the level of the technologies of participating companies are therefore not a great problem, and conflicts of interest do not rise to the fore.

In the past, collective research was carried out in the form of a voluntary organization. However, in this type of organization the responsibility for asset management and accounting was not clear, hence no tax incentive could be given for collective research even if social benefits were produced.

The main reason for this inconvenience was that the organization for collective research did not have the status of a legal corporation. Therefore, in order to solve this problem, the Engineering Research Association Act was enacted on May 1961. Under this law, the collective research activity designated by the government as an ERA can receive the following tax benefits:

1. Tax deduction of expenditures charged by cooperatives to their members for the membership fee and the acquisition of fixed assets for research;
2. Condensed recording, as small as one yen, of charges imposed by cooperatives for the acquisition of fixed assets required for the study of technology; and,
3. Accelerated depreciation of fixed assets for research up to two-thirds in the first three years.

5.2 Historical development of collective research

By 1985, a total of seventy-six ERAs were established, fifty of which were active as of July 1985. In other words, by 1985, twenty-four ERAs had already completed their missions and so had been disbanded. [10] The time trend in the number of ERAs established in each year is shown in Table 4.1.

In October 1961, the first ERA was established to conduct research in polymer materials technology. In the 1960s, the emphasis was placed on bringing Japanese technology up to the world level. A typical example for this era was the ERA in the optical industry. Around the mid-1970s, the emphasis shifted from catching up to generating basic technologies and to a stronger relation with long-term national research programs. The epoch-making example of this kind of shift was the ERA for VLSI development which was established in March 1976.

Because MITI's policy innovation so far has been an unconscious part of catching up with the advanced countries, the Japanese development of the ERAs was not, of course, free from mistakes. There is much evidence

Table 4.1 Time-series of ERAs established in each calender year

Year	No. of ERAs	Year	No. of ERAs	Year	No. of ERAs
1961	4	1971	2	1981	7
1962	5	1972	3	1982	9
1963	0	1973	8	1983	11
1964	4	1974	3	1984	5
1965	0	1975	4	1985	5
1966	0	1976	3	1986	12
1967	0	1977	2	1987	6
1968	0	1978	2		
1969	0	1979	3		
1970	1	1980	5		

Source: I. Shirai and F. Kodama, 'Quantitative analysis on structure of collective R&D programs by private corporations in Japan', p. 4, NISTEP Report no.5, August 1989.

to illustrate MITI's trial-and-error process. [10]

A problem of R&D management occurred when MITI launched the National R&D Program (popularly known as the large-scale project) in 1966. This was the first attempt by the Japanese government to finance 100 percent of the R&D carried out by private enterprises. The model for this proposal was the defense R&D programs in the United States and European countries.

At this point, MITI assumed that a National R&D Program was incompatible with the ERA scheme, and as is shown in Table 4.2, no new ERAs were established from 1966 to 1970, while as many as five Large-Scale Projects were launched during this period. In other words, the ERA scheme was not used as a vehicle to implement the Large-Scale Projects.

Table 4.2 Compatibility between ERA and Large-Scale Project

Year	New ERA	New LSP
1965	2	–
1966	0	2
1967	0	1
1968	0	0
1969	0	1
1970	0	1
1971	3	3
1972	3	0
1973	2	3
1974	9	0

However, it was discovered that the Large-Scale Project scheme was compatible with the ERA scheme and that they complement each other in some ares of research. It has since become normal practice to use ERAs for the implementation of MITI's R&D programs.

6. Technological rationale of collective research

As to the technological rationale, our analysis is concentrated on special cases in which the scheme of a 'cooperative game' holds. The organizational problem in the cooperative game is how to create and guarantee a situation in which the cooperative game is to be conducted among rival firms. This is clearly related to the problems of policy formulation and of the management of research organizations.

Functionally, however, collective action through cooperative games creates demand articulation for technology. Therefore, our analysis of the technological rationale behind collective research by rival firms centers around the concept of 'demand articulation'. [7] Through this process of demand articulation, the need for a specific technology manifests itself and the R&D effort is targeted toward developing and perfecting it.

However, this concept of demand articulation has developed for an analytic framework of technology development processes conducted by a single firm. Therefore, we can claim that the dynamic process of collective action by rival firms creates the functional equivalent of demand articulation in a single firm. In other words, it can be conceptualized as 'collective articulation of demand'.

The demand articulation promoted by collective research should be viewed in and can be explained by the overall framework of industrial technological linkages. It can assist in creating a national technological infrastructure. Sometimes it results in establishing upstream technological linkages, and at other times it contributes to establishing downstream linkages.

As far as the methodology of research is concerned, the case study method is effective, because successful collective research projects in which the cooperative-game scheme holds are more or less special cases. Therefore, in this section, the two prototypical examples of Engineering Research Associations are selected for case studies of the cooperative game. Specifically, the examples are the ERAs for the development of very large scale integration and the development of pattern information processing system.

6.1 Prototypical examples

DEVELOPMENT OF VERY LARGE SCALE INTEGRATION

The Engineering Research Association for VLSI existed from 1976 to 1979 and spent a total of ¥73.7 billion, of which ¥29.1 billion was paid by the government on a project funding basis. Members of the association were: Fujitsu, Hitachi, Mitsubishi, NEC and Toshiba. These five companies established a joint research laboratory within the association.

This joint research laboratory had around 100 researchers who were sent from these member companies and from the Electro-Technical Laboratory (ETL), one of the National Research Institutes. Approximately 20 percent of the research was carried out in this joint research laboratory and the remaining 80 percent was done by the individual member companies in their own laboratories, with a steering committee in the ERA as a coordinating body.

The research was divided into the following six areas: (1) micro-manufacturing technology, with an emphasis on the development of the equipment; (2) crystal technology, with an emphasis on the analysis of the crystallizing process of silicon; (3) product-design technology; (4) process technology; (5) testing and evaluation technology; and (6) devices technology. All the research items related to (1) and (2) were carried out in the joint research laboratory by a team composed of researchers from the different organizations, while those related to (3) were carried out by the individual member companies in their own laboratories. As far as the items related to (4)–(6) are concerned, only the portions which were basic by nature and common to all the members, were carried out in the joint research laboratory. [11]

A large percentage of the research and development carried out in the joint research laboratory was subcontracted to the supplier companies, which were not members of the association. Those suppliers which were heavily involved were: Nikon, which developed the optical wafer stepper; JOEL, which developed electron beam lithography; printing companies, which developed lithography; and silicon crystal suppliers.

Pattern information processing system

The R&D project on pattern information processing system (PIPS) was launched in 1971 as one of MITI's Large-Scale Projects, under the guidance given by the ETL. As many as ten companies participated in this project and as many as eighteen research items were investigated with ETL responsible at least for some parts of all of them. This project continued until 1980 when total funding reached ¥22 billion. The project consisted of the following four research areas: (1) pattern recognition systems; (2) devices and materials; (3) information processing systems; and (4) an integrated system prototype. [12]

Those research items related to research areas (1)–(3) were contracted directly to the ten companies, while those related to (4) were contracted to an ERA, which consisted of five companies: Hitachi, Toshiba, NEC, Fujitsu and Mitsubishi Electric. This ERA came into existence as a voluntary organization in 1975, five years after the Large Scale Project was launched, and as a legal corporation in 1977. It functioned as a coordinating body for the companies whose hardware architectures are different from one another and facilitated information exchange for integration into a prototype system.

The system is composed of the following six recognition subsystems: printed *kanji* (Chinese character) recognition; handwritten recognition; gray picture recognition; color picture recognition; object recognition system; and speech recognition. Each firm was responsible for one of these subsystems. Therefore, the research on each subsystems was carried out rather independently and thus they were not as integrated as had been expected.

However, the project yielded several component technologies, such as *kanji* processing and image processing in the field of software; a high-speed reader of handwritten census data, an X-ray TV camera and remote sensor in the development of equipment; and a magnetic bubble memory in the field of hardware.

6.1 Rationale for upstream technological linkages

In joint research by rival firms in the same industry, success hinges largely on ensuring that the research is basic and of common interest to all the participants. In the VLSI research association, therefore, rather than focusing on the method of producing chips, research efforts centered around developing a prototype for IC manufacturing equipment and analyzing a process for the crystallization of silicon, a basic material in chip production. More importantly, no manufacturers of either production equipment or chip materials were among the participants.

In Figure 4.1, major actors involved in the Japanese development of VLSI and the technical linkages among them are depicted. [13] The specific activities of the association included the development of the optical stepper. The research laboratory of lithography sought to reduce the electronic circuit onto the silicon base optically, not electronically.

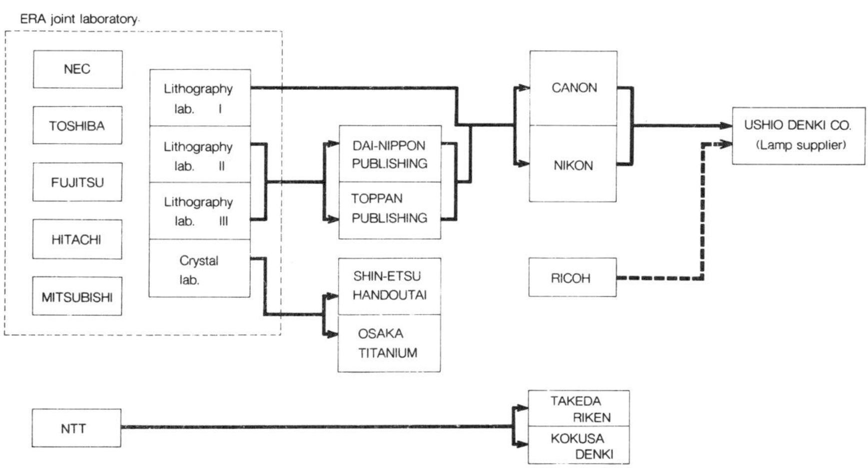

Figure 4.1 Upstream technical linkages of Japanese VLSI development

Therefore, this laboratory contracted research to camera manufacturers who owned the lens technology and thus companies such as Nikon and Canon succeeded in developing the optical stepper. [13]

At the same time, the results of the silicon crystallization process research were passed along to the Shin-Etsu Semiconductor Company and the Osaka Titanium Compnay, both of which came to comprise Japan's silicon production industry. Table 4.3 shows the Japanese companies' shares in the world production of semiconductor, fabrication equipment, and materials.

Table 4.3 World market share of Japanese production of semiconductor-related products

Year	Semiconductor	Fabrication equipment	Materials
1980	27%	18%	na
1981	34	22	na
1982	36	28	na
1983	39	30	na
1984	40	32	na
1985	42	34	60%
1986	48	34	68
1987	48	36	71
1988	51	47	71
1989	51	49	71

Note: Equipments do not include testers. na means not available.

Source: Dataquest.

Although their share of semiconductor production was 51 percent of the world total in 1989, it was 49 percent in fabrication equipments excluding testers, and reached as high as 71 percent in materials production. A closer look at the changes since 1980 clearly reveals that the production share of fabrication equipment has caught up quickly, but that it did not reach that of semiconductor until after 1987. It was not easy for the equipment suppliers to penetrate foreign markets as it was necessary to provide good technical service to the users. [13] On the other hand, the share of materials had already by far surpassed that of semiconductor. In other words, if you go up toward upstream sectors, the Japanese dominance in the world production becomes more obvious.

Therefore, we can conclude, by getting all the major chip manufacturers together in one place, the articulation of demand for manufacturing equipment and materials for chip-making was implemented. In this way, the infrastructure was established through the efforts of demand articulation by the chip manufacturers. [10]

However, in order to perfect demand articulation for chip-making, we must investigate the following items: second-tier suppliers and testing machines. [13] As far as the multiple nature of the supplier is concerned, the suppliers of optical steppers, i.e. first-tier suppliers, are not the only beneficiaries of this joint effort. We can find the real beneficiary within the second-tier suppliers. As shown in Figure 4.1, Ushio Denki, the supplier of the lamp used for the optical stepper, ended up dominating the world market for lamps much more than the Japanese suppliers of optical stepper. Ushio had a market share of 100 percent for aligner lamps in Japan, and 50 percent of the global market in 1983.

As far as the testing machine is concerned, the VLSI Research Association did not pay enough attention to its importance. Thus it did not work as a demand articulation mechanism for the tester. What made demand articulation for the testing machine possible was the joint research managed by the telecommunication company, NTT. In 1977–81, NTT conducted joint research on the next generation of testers with the Takeda Riken Company, which became the major supplier of this type of testing machine in the world market. Takeda Riken's worldwide market share is 20 per cent. However, when it comes to memory testers, Takeda is clearly the leader with 30 percent of the global market in 1983.

In this joint research, detailed requirements for the new tester were collected from major VLSI manufacturers such as NEC, Fujitsu, Hitachi and Oki. However, the fundamental requirements were eventually set by NTT after several meetings to work out the joint specifications.

Through the ten years' demand articulation efforts initiated by the VLSI research association, the Japanese companies in the upstream sector of chip manufacturing are beginning to emerge as dominant players in world production. The changes in the world top ten suppliers of semiconductor manufacturing equipment are shown in Table 4.4.

As far as the structural aspects of demand articulation are concerned, the multi-level and multi-pole hierarchical structure exists in the causal chain

Table 4.4 The world top ten equipment manufacturers

	1980	1985	1989
1	Perkin Elmer	Perkin Elmer	*Tokyo Electron*
2	GCA	*Tokyo Electron*	*Nikon*
3	Applied Material	General Signal	Applied Material
4	Sclumberger	Varian	*Takeda Riken*
5	Varian	Teradyne	*Canon*
6	Teradyne	Eaton	General Signal
7	Eaton	Schlumberger	Varian
8	General Signal	*Takeda Riken*	*Hitachi*
9	Kulicke Soffa	Applied Material	Teradyne
10	*Takeda Riken*	GCA	ASM International

Note: Japanese companies are in italics.

Source: VLSI Research Inc.

of demand articulation. In terms of multiplicity of level, there exists a manufacturer level, a component supplier level and a sub-component sub-supplier level. In terms of multi-polarity, MITI's project articulated the demand for optical steppers and materials, while NTT's project did the same for the testing machine.

6.3 Rationale for downstream technological linkages

As described before, MITI assumed that the Large-Scale Project was incompatible with the ERA scheme. In fact, there was no new ERA established for four years after the Large-Scale Project system was implemented. In the case of the PIPS project, however, it was found that the Large-Scale Project scheme was compatible with the ERA scheme and that they complemented each other in some areas of research. It then became normal practice to use ERAs for the implementation of MITI's R&D programs.

There seem to be two reasons why the two schemes are compatible: one is administrative and the other is technological. When the government directly contracts firms, all of the detailed administrative work must be performed within the government, which is not well prepared for this kind of work organizationally or technically. In order to perform the work smoothly, the government has to hire extra staff, whose status is supposed to be that of a government employee.

In terms of the technological reasons, the two schemes become compatible when it comes to an R&D project that involves system integration, such as the PIPS project. In this type of project, the government assigns the contract for various component technologies to different companies. Therefore, some kind of collaborative effort is needed among the participating firms at the system's comprehensive design stage, and during the final testing and evaluation of the integrated system. However, the government, or the participating firm, is not the appropriate place for such coordination. Therefore, the ERA becomes necessary as a vehicle for this work.

As described before, the PIPS project was launched in 1971 as a Large-Scale Project. However, a research association had to be established in 1977 to conduct research on an integrated system prototype. It worked as a coordinating body for the companies whose hardware architectures differed from each other so that they could exchange information in order to integrate a prototype system.

One of the major issues in national technology policy is how to manage the science–technology interaction, particularly the interaction between the national research laboratories and industry. An ERA can be a channel for transforming scientific ideas – mainly developed in research laboratories into technical innovations – which is the major responsibility of industry.

In many cases, the research director of an association is recruited from a relevant national research laboratory because he has the most advanced knowledge and can be supposed to be neutral among the vested interests of the participating firms. This position of neutrality becomes critical when conflicts of interest occur at the stage of integration, such as in the PIPS project. Someone has to play the role of referee.

Moreover, researchers at national research laboratories are heavily involved in the planning work for the association and make substantial contributions to setting project goals. In the PIPS project, as described before, the system is composed of six recognition subsystems. Each firm was responsible for one of these subsystems. At one time, the participating firms were reluctant to conduct research because they thought the goals set for each subsystem were unfeasible. But they were persuaded to do so because researchers at the ETL had already conducted some basic research indicating the feasibility of the goals that had been set. [14]

7. Database for studies of collective research

As the significance of Japan's high-tech advance began to sink in around the world, many outside observers began to entertain the suspicion that Japan's research associations, which allow joint research by rival firms, might amount to a conspiracy between the government and the business community. [15]

Success in joint research by rival firms in the same industry requires success in both forming the association and actually conducting the research. [16] However, scientific and quantitative analysis can be applied only to the first stage. The judgment concerning the success of the quality of the joint research is at least subjective, and its impact can be analyzed only several years after joint research is disbanded. [17] Therefore, in what will follow, our analysis is focused on the participating behavior of rival firms. And we will try to show that the rival firms' participation in collective research is a rational behavior in terms of economics.

In order to analyze rival firms' economic incentives for participation in collective research, we first need a classification scheme. One obvious taxonomic question is whether the joint effort is centered around a science project or a technology development project. The objective of the former type is to gain the scientific knowledge, thus the concept of 'public goods' can be applied. On the other hand, the objective of the latter is to acquire technical know-how, which can be commercialized later by the participating firms. Therefore, the public goods concept is no longer effective for the analysis of technology development projects.

The other taxonomic question is whether the joint effort is conducted on an international basis or on a domestic basis. The examples of the former are internationally collaborative Antarctic expeditions and global climate research. So far, the majority of international collaborations are scientific projects with a few exceptions. The majority of collective research on technology developments is conducted within national borders. Therefore, an ideal demarcation of collective research is between the following two categories: one which deals with science and is conducted on an international basis; and the other which deals with technology development and is conducted within one country.

7.1 International collaboration in science

As typical examples of international collaboration, international joint projects by the IEA are chosen. The IEA was established in November 1974, based on a recommendation by the OECD (Organizations for Economic Cooperation and Development), and its membership covers twenty-four countries. It was established as an autonomous institution within the OECD framework. Its objective is to develop both conventional and alternative energy technologies. Its major activity is to create and to manage international collaboration in energy-related R&D projects. Among other things, participation of at least three countries and a limited time frame are required as a rule. The project proposal is reviewed by the working party, which nominates a lead country. This proposal is then distributed to all the member countries to ask for participation. Each country independently decides on its level of participation.

As of 1984, there were sixty-one projects [18] which cover the following subjects with the number of projects so far implemented under each subject category in parenthesis:

systems analysis	(1)
energy use	(24)
fossil fuels	(9)
geothermal energy	(2)
solar energy	(9)
bio-mass energy	(2)
ocean energy	(1)
wind energy	(6)
hydrogen energy	(3)
nuclear fusion	(4)

However, the systems analysis project is the first project of the IEA, and is very general. It has the most participants – as many as eighteen countries. Therefore, this project is excluded from the sample, because it is so different from the other projects in its nature and size and we analyze only the remaining sixty projects.

Participation also varies from country to country. Some countries such as the United states and Sweden participated in 70 percent of all the projects, while others such as Mexico and Finland have participated in only one project so far. Therefore, only those countries who have participated in more than two projects as a lead country were selected for the final database. Thus we have ten countries which satisfy these conditions.

The frequency distribution of the number of these ten countries that participated in the sixty projects is displayed in Table 4.5. The frequency distribution curve is depicted in Figure 4.2. As can be seen in the figure, it is bell-shaped. It looks like a binomial distribution, which alludes to the fact that the decision-making is conducted independently among the countries.

Table 4.5 Frequency of number of participants in an IEA Project

Number of participants	Observed frequency
1	1
2	2
3	11
4	13
5	10
6	10
7	4
8	6
9	2
10	1
total	60

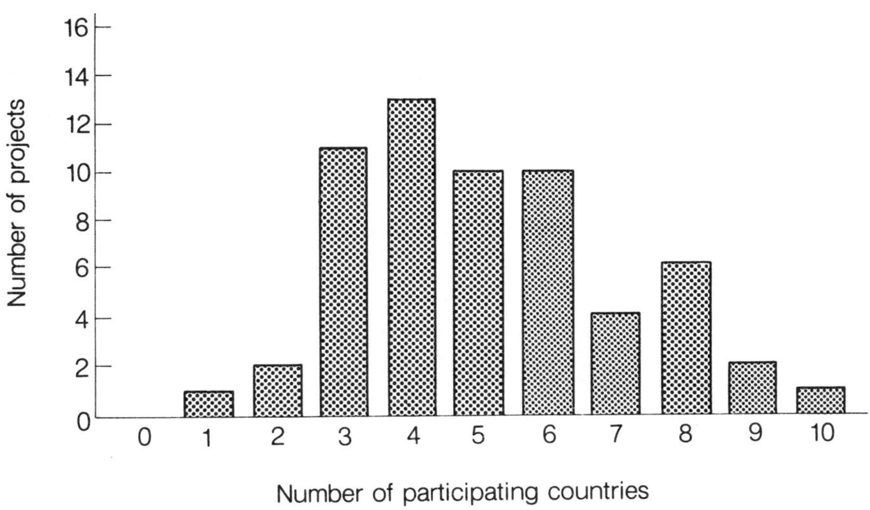

Figure 4.2 Frequency distribution of a number of participants in an IEA project

7.2 Number of participating firms in a research association

As far as the fifty ERAs which existed in 1985 are concerned, the number of participating firms has a frequency distribution shown in Figure 4.3. The number is distributed in a wide range from three to as many as forty-two, which is the case of Technology Research Association of Medical and Welfare Apparatus. We cannot find any regularity in this distribution curve. One might say that it follows approximately a bell-shaped normal distribution. However, all that we can say is that this distribution occurs because of the central limit theorem: the samples are drawn from several populations of different categories.

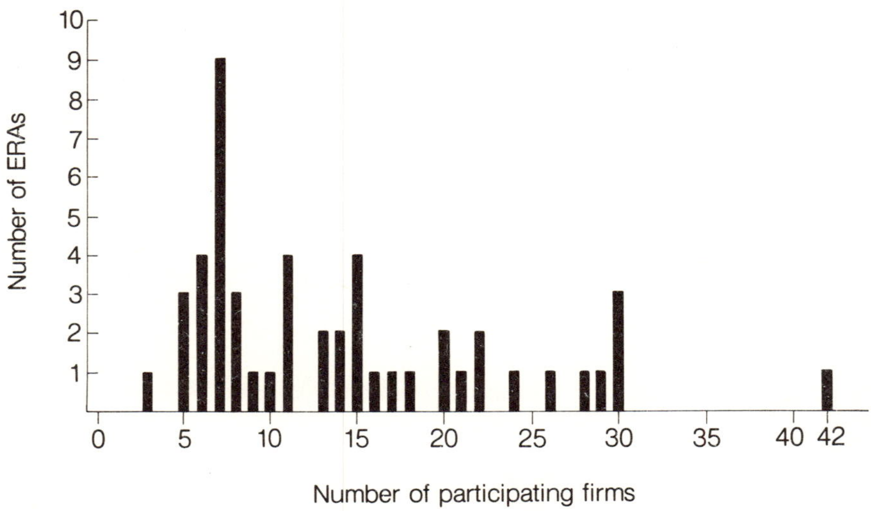

Figure 4.3 Frequency distribution of number of participating firms in an ERA (in ERAs active at 1985)

We are interested in collective action among competing firms. Therefore, only associations composed of rival firms are selected. Those industrial sectors in which more than three companies exist and in which those companies are similar in size, are selected for study. [19] This includes steel companies, textile companies, computer companies and shipbuilding companies, as shown in Table 4.6.

The frequency distribution of the number of participating firms is shown in Table 4.7. In order to expand the database, those collective researches organized by the Japan Key Technology Center, which was established in 1985 as an extended version of ERA, are included together with research associations.

Table 4.6 Names of rival firms selected for analysis

Steel companies	Textile companies	Computer companies	Shipbuilding companies
Shin-Nihon	Toray	Toshiba	Mistubishi
Nihon-Kohkan	Teijin	Hitachi	Kawasaki
Kawasaki	Asahi-Kasei	Fujitsu	Ishihari
Kobe Steel		Mitsubishi	
Sumitomo		NEC	

Table 4.7 Frequency distribution of participation by selected rival companies (as of 1988)

Industrial sector	Number of projects	Number of firms participating				
		1	2	3	4	5
steel	40	27	1	1	4	7
textile	17	11	2	4	–	–
computer	67	27	15	6	4	15
shipbuilding	37	17	9	11	–	–

Source: I. Shirai and F. Kodama, 'Quantitative analysis on structure of collective R&D programs by private corporations in Japan', NISTEP (National Institute of Science and Technology Policy) Report no.5, 1989. [19]

As seen in Table 4.7, there are as many as sixty-seven cases of collective research in which any one of the rival computer manufacturers participated: forty cases in the steel companies, thirty-seven in the shipbuilding companies and seventeen in the textile companies. The distribution curve of how many of these companies participated indicates that in all the industrial sectors the pattern of participation is U-shaped, almost opposite to the pattern observed in Figure 4.2. The difference in the observed phenomenon alludes to the difference in the decision-making of participation.

7.3 Joint research by equipment manufacturers with NTT

Although the pattern of participation of rival firms in ERAs alludes to the differences in decision-making, the sample is not large enough for statistical inference. Therefore, we need other sources for a database.

There are a substantial number of joint research projects by communication equipment manufacturers with NTT, which used to be a public company. In fact, there are a lot of similarities between the joint research with NTT and that in an ERA. Before the privatisation of NTT, its function had been similar to that of the government. The joint research organized by NTT is of course to serve the development of technologies utilized by NTT. And its joint research is among rival firms, as often the case in an ERA. Moreover, since NTT is prohibited by law from owning a manufacturing function it has developed stable relations with the equipment and cable manufacturers.

Although the number of joint research projects by NTT with equipment manufacturers is not known, we can estimate it by counting the joint application of patents [10] although we cannot generally assume that joint patent application is a result of joint research. However, it is hard to imagine joint patent applications not having joint research and single applications being the result of joint research. We cannot also assume a one-to-one correspondence between a joint application and joint research, because one joint research project might produce several patent applications. However, we are concerned with the participating pattern in joint research. Therefore, we can safely assume that the statistics of joint application reflects the participation pattern of joint research.

In 1986, as many as 655 joint patents were applied for by NTT jointly with one or more of the four equipment manufacturers. The time-series data of the joint application of patents is shown in Table 4.8.

Table 4.8 Number of joint patent application with NTT

Year	Total number of patents jointly applied	Number of manufacturers involved in an application			
		1	2	3	4
1980	627	457	50	7	49
1981	669	426	59	20	164
1982	652	479	33	8	132
1983	655	487	49	25	94
1984	540	374	28	5	133
1985	375	270	18	15	72
1986	655	532	53	22	48

As is shown in Table 4.8, NTT cooperates most frequently with only one equipment manufacturer, cooperation with all the four equipment manufacturers is second most frequent, and cooperation with two or three manufacturers is less frequent. Although the number of cases involved is substantial, this pattern of participation is quite stable over all the years, as shown in the Table.

However, in order to avoid yearly fluctuations, three-year moving averages were calculated, and used for the analysis in the following sections. Its relative frequency distribution curve is depicted in Figure 4.4. As is shown in the Figure, roughly speaking, the cooperation with only one manufacturer accounts for 60–70 per cent, and that with all four manufacturers is 15–20 per cent, while that with two manufacturers and with three manufacturers accounts for 8–10 percent and 2–3 percent respectively.

It is an obvious manifestation of the U-shaped distribution, to which the observed phenomena of rival firms' participation in the ERA alludes. It also indicates clearly that the decision-making on participation is very much interdependent.

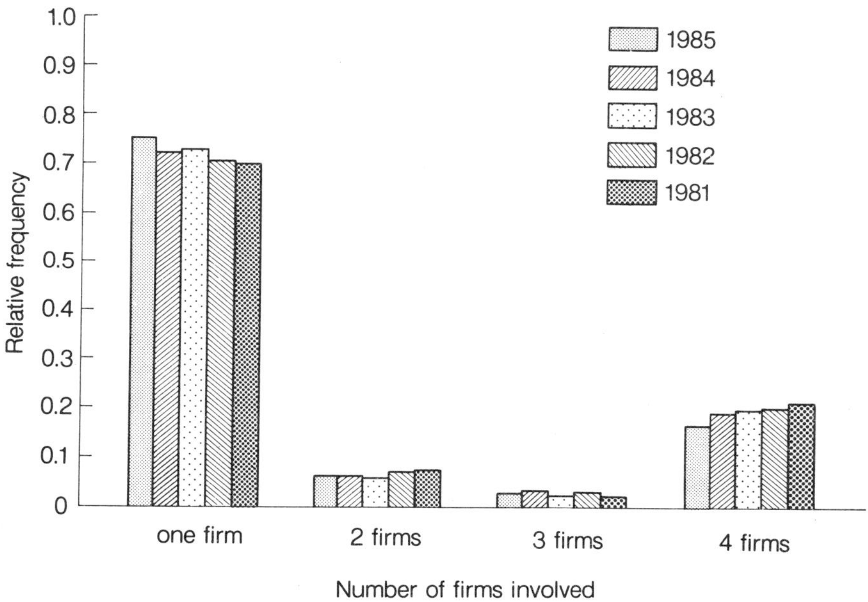

Figure 4.4 Frequency distribution of a number of equipment manufacturers involved in a joint patent application with NTT (three year moving average)

8. Economic rationale of participation

Our analysis of economic rationale is centered on whether the decisions about participation are independent or dependent. For this purpose, several mathematical models are to be constructed based on the various assumptions of decision structures. Statistical tests are conducted to determine whether models can explain the observed phenomena of the number of participating agents. [20][21]

8.1 Cooperation in science: independent decision models

In the case of international collaboration in science, it is observed that the frequency distribution of the number of participating countries shows a bell-shaped distribution, as illustrated in the IEA programs. If we assume that each country makes its decision independently from each other, we can think of the following two types of models: one is a static model; and the other is a dynamic model.

STATIC INTERPRETATION: BINOMIAL MODEL

Let p be the probability that a country participates on a project, then the probability distribution of the random variable X, i.e. the number of participating countries, follows a binomial distribution expressed mathematically by:

$$Pr(X = k) = {}_{10}C_k \cdot p^k \cdot (1-p)^{n-k},$$

where k is the number of participants, and n is the total possible number of countries.

By letting the participating probability, p, be such that the observed average coincides with the calculated average, i.e. $n \cdot p = 5.07$, and by letting $n = 10$, the probability distribution can be estimated. The difference between the observed value and the calculated value is shown in Table 4.9. It looks as if a binomial distribution could explain the observed values, although the X-square is 26.38, a little bit above 18.31, the 5 percent point for a X-square distribution with 10 degrees of freedom. However, we can think of various ways to modify the theoretical models, in order to improve the fit.

Table 4.9 Difference between observation and binomial model

Number of participants	Observed frequency	Binomial model
0	0	0.05
1	1	0.53
2	2	2.37
3	11	6.66
4	13	11.97
5	10	14.75
6	10	12.63
7	4	7.41
8	6	2.85
9	2	0.01
10	1	0.00
X-square value		26.38

DYNAMIC INTERPRETATION: POISSON MODEL

One method is to formulate the participation process by means of the 'Poisson' process. Poisson distribution can be derived as the distribution of the number of random arrivals during a limited time interval, if arrival phenomena are stochastically independent of each other. The analogy is a statistical identity between the random arrival of countries in a specified time period for application and the participation of countries whose decision is made independently.

Let μ be the average number of arrivals in a given interval, then the probability that the number of arrivals is k, $Pr(k)$ is described by:

$$Pr(k) = \mu^k \cdot e^{-\mu}/k!.$$

By letting μ be the observed average, i.e. $\mu = 5.07$, the probability distribution can be estimated. The difference between the observed value and the calculated value is shown in Table 4.10. As can be seen in the table, X-square value is 6.30, far below 18.31, the 5 percent point for X-square distribution with 10 degrees of freedom.

Table 4.10 Difference between observation and Poisson model

Number of participants	Observed frequency	Poisson model
0	0	0.38
1	1	1.92
2	2	4.85
3	11	8.20
4	13	10.38
5	10	10.52
6	10	8.89
7	4	6.43
8	6	4.07
9	2	2.68
10	1	1.16
X-square value		6.30

Therefore, we can conclude that the participation in international science projects is more or less decided by each country on its own, namely, independent of each other. This result indicates in a very indirect manner that research outputs of international collaboration are more or less public goods, which do not displace their utilization by non-participants. One of the policy implications is as follows: in science-related projects, an intuitive judgment is valid, and one in which a certain number of participants somewhere between one participant and all the possible participants are to be expected, with a maximum probability of participation around the average.

8.2 Cooperation in technology: interdependent decision models

The distribution of the number of participating firms in domestic cooperation on technology development has a U-shape, as is shown clearly in the NTT-initiated joint research with the four equipment manufacturers. A similar phenomena is observed in the research associations of rival firms in major industries. All these indicate that the decision is based on considerations of both cost and benefit.

Problem formulation by non-cooperative game

In this situation, the participation behavior can be formulated within the framework of a 'non-cooperative game',([22]–[26]) where the four equipment manufacturers are players. The rules of the game are as follows:

1. NTT asks the four players one by one their intention to participate in the game. The game is terminated after NTT asks all the four players.
2. Each player is informed on the benefit and the cost of all the possible combinations of participation. And they are also informed of the penalty of non-participation, while the other players participate.

Each player then behaves so as to maximize his pay-off, i.e. the benefit minus cost minus penalty. The game theory tells us that each player decides to participate only if the benefit exceeds the sum of the cost and the penalty. Therefore, if we know the rule by which benefits and costs are allocated, we can formulate the decision structure of participation.

The rule of allocation of benefit and cost

Although there are many variations of allocation rules, let us assume the following simplest set of rules:

1. The total cost of the project is independent of the number of participants. Therefore it is denoted by c.
2. The total benefit of the project is independent of the participation pattern, and is denoted by b. It is allocated equally to all the participating players.
3. The cost c is shared equally by NTT and participants.
4. The penalty of non-participation is independent of the participation pattern. It is denoted by d.

The pay-off of the player can then be derived. Let n be the number of participating player, the pay-off $U(n)$ can be described as:

$$U(n) = b/n - c/(n+1) - d.$$

Let $z = b/c$ (the benefit/cost ratio), then

$$U(n) = c\cdot[z/n - 1/(n+1)] - d.$$

Let us assume that $d = 0$, for simplicity, then

$$U(1) = c\cdot(z - \tfrac{1}{2}),$$
$$U(2) = c\cdot(\tfrac{z}{2} - \tfrac{1}{3}),$$
$$U(3) = c\cdot(\tfrac{z}{3} - \tfrac{1}{4}),$$
$$U(4) = c\cdot(\tfrac{z}{4} - \tfrac{1}{5}).$$

PARTICIPATION CONDITIONS

It is reasonable to assume that each player will make a decision to participate if the pay-off is positive, and decide not to participate if the pay-off is negative. Therefore, the participation condition can be derived as follows:

if $0 < z < \frac{1}{2}$,	then no firm participates;
if $\frac{1}{2} < z < \frac{2}{3}$,	then one firm participates;
if $\frac{2}{3} < z < \frac{3}{4}$,	then two firms participate;
if $\frac{3}{4} < z < \frac{4}{5}$,	then three firms participate;
if $\frac{4}{5} < z$,	then all the four firms participate.

In the case where z is less than $1/2$, there is no participation if we follow the rule of equal cost-sharing. However, if NTT is ready to share more than half the cost, it is possible for one firm to participate. If NTT shares more than $\pi = (c-b)$, then the pay-off becomes positive for one firm. The alternative for NTT is to do it alone, carrying the full cost. It is reasonable to assume that NTT is ready to share the cost of $\pi(=c-b)$, and therefore, even if z is less than $1/2$, we can assume that one firm will participate. Then, we can assume, one firm will participate, if $0 < z < 1/2$.

On the other hand, the situation becomes quite different if z is larger than 1. In this case, the benefit exceeds the cost without any sharing. In other words, it pays off if any manufacturer, not just one of the four equipment manufacturers, develops the technology alone. Therefore, NTT has no reason to initiate joint research, because it will receive the benefits generated by the in-house research of the manufacturer. The same situation applies to any one of the four manufacturers. Therefore, we can assume that it is not possible for joint research to be realized. Thus all the four firms will participate only if $4/5 < z < 1$.

DERIVATION OF PARTICIPATION FREQUENCY

All the findings described above can be integrated in Figure 4.5, where the benefit/cost ratio, z, is plotted along the X-axis and the cost shared by NTT, π, is plotted along the Y-axis. As illustrated by the figure, it is shown that the number of participants is uniquely determined by the value z of benefit/cost ratio. Therefore, we can derive the probability distribution of participating firms, if the probability of z is given.

Let $f(z)$ be the probability density function of z, then the probability that the number of participants is n, $q(n)$, is derived by the following formula:

$$q(1) = \int_{z=1/2}^{z=2/3} f(z)\cdot dz + \int_{z=0}^{z=1/2} f(z)\cdot dz,$$

$$q(n) = \int_{z=n/(n+1)}^{z=(n+1)/(n+2)} f(z)\cdot dz, \qquad \text{for } n = 2, \text{ or } 3,$$

$$q(4) = \int_{z=4/5}^{z=1} f(z)\cdot dz.$$

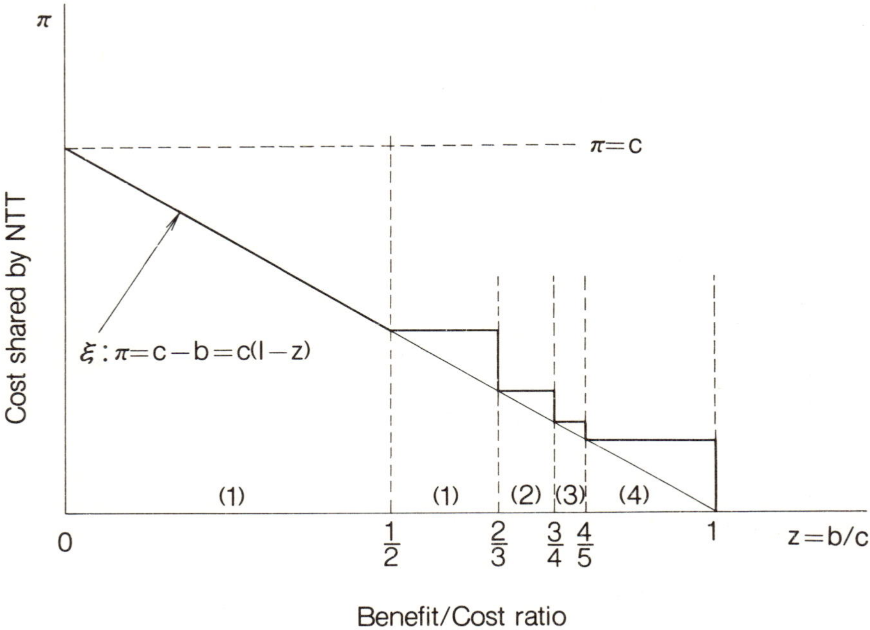

Figure 4.5 Relation between benefit/cost ratio and cost shared by NTT

CALCULATED FREQUENCY

There is no relevant information concerning the mathematical form of the probability density function of z. Therefore, the uniform distribution is selected tentatively because it is the simplest form mathematically. [20] The distribution is then proportional to the length of intervals between the boundary values, as can be seen in Figure 4.5. The result is as follows (Table 4.11):

Table 4.11 Difference between observation and values calculated by uniform distribution of *z*.

Participation pattern	Calculated	Observed
one-firm participation:	66.7%	70.4%
two-firm participation:	8.3	7.1
three-firm participation:	5.0	2.7
four-firm participation:	20.0	19.7

Note: The observed are three-year moving averages at 1982.

Therefore, even by the simplest assumption, this model can regenerate the basic characteristics concerning the distribution of participating firms: where one firm's participation is the most frequent, four firms' participation is the second frequent, two firms' participation is the third and three firms' participation is the least probable.

Therefore, we can say that this model can simulate the basic characteristics of the frequency distribution of participation behavior, at least to the first order. Thus we can conclude that participation in a technology project is based on dependent decision-making.

However, if a strict significance test is applied, it yields the following *X*-square values:

15.42	for 1981;
9.53	for 1982;
17.94	for 1983;
10.41	for 1984;
26.09	for 1985.

We cannot say that the model fits the observations statistically, because the 5 percent point is 7.78 with 3 degrees of freedom. However, there are several ways to improve the model's performance.

Modification of the model

First of all, we can modify the assumption that $d=0$. It is reasonable to assume that the larger the cost c is, the larger is the penalty of non-participation d. Therefore, we can assume that d is proportional to c. By letting $d/c = 0.02$, for example, we can reduce the probability of two firms' participation to 6.3 percent and that of three firms' participation to 3.0 percent, although that of four firms' participation increases by 6 per cent.

A modification can be also made of the assumption that the benefit/cost ratio of z is uniformly distributed. By considering the necessity of collective research, it is reasonable to assume that its distribution among a whole population of collective researches skewed toward smaller values between 0 and 1. Therefore, we can modify the uniform distribution of z, to a decreasing function of z, for example, such as $f(z) = 1.3 - 0.6 \cdot z$. Then, the following estimation is available (Table 4.12):

Table 4.12 Difference between observation and values calculated with a decreasing function of *z*

Participation pattern	Calculated	Observed
one firm	71.5%	70.4%
two firms	5.7	7.1
three firms	2.6	2.7
four firms	20.2	19.7

Note: The observed are three-year moving averages at 1982.

It is needless to say that the statistical test of significance is no problem at all. A statistical test, for example, yields the following *X*-square values:

4.88	for 1981;
2.57	for 1982;
1.03	for 1983;
0.62	for 1984;
8.28	for 1985.

As can be seen above, the *X*-square values for all the years except 1985 are far below 7.78, the 5 percent point with 3 degrees of freedom. Therefore, we can conclude that the decision to participate in a domestic technology development project is interdependent among rival firms. However, we can also conclude that participation is economically rational behavior, which is strictly based on estimates of benefit/cost ratios. Therefore, we can conclude, hypothesis of collusion or 'conspiracy' are not needed to explain corporate cooperative research even in applied research or development. [15]

9. Derived dichotomy of collective research

Through both quantitative and descriptive analyses, we can show that the participation of rival firms in collective research has both an economic and a technological rationale. The results of analysis can be summarized as shown in Table 4.13.

Table 4.13 Derived dichotomies of collective research

Science	vs	Technology		
bell-shaped public international independent intuitive		U-shaped distribution private good domestic program dependent decision counter-intuitive		
		Economic rationale	vs	Technological
		non-cooperative game average cases random sample quantitative analysis mathematical modeling		cooperative special successful descriptive case study
IEA projects		NTT-initiated research ERAs among rival firms		ERA for VLSI ERA for PIPS

In terms of the economic rationale, the demarcation between a science project and a technological project is important. In a science project, decision-making regarding participation is made independently among the possible participants. This results in the statistical phenomena in which the probability distribution of the number of participants in any single project is bell shaped. In a technology development project, on the other hand, decision-making is interdependent among rival firms. This produces the statistical phenomenon of a U-shaped distribution of the number of participants in any one project.

This U-shaped phenomena can be explained as the result of economically rational behavior by rival firms, in which the decision to participate is made on the basis of each firm's estimate of the benefit/cost ratio. In other words, rational decisions made by rival firms lead to seemingly counter-intuitive collective behavior, in which participation by one firm or participation by all rival firms is more plausible than the cases in-between.

All these findings provide insights that are useful with regard to policy formulation for collective research. In science-related projects, an intuitive judgment is valid; one can expect a certain number of participants somewhere between one participant and all the possible participants, with a maximum probability of participation around the average. In technology development projects, on the other hand, such an intuitive judgment is misleading, because the two extreme cases are more plausible than the cases where somewhere around half of all the possible candidates are to participate. However, it is still unknown whether this counter-intuitive collective behavior can occur in an international collaboration on technology development.

The technological rationale is important in special cases in which the firms are involved in a 'cooperative game'. This was seen in the case studies of the VLSI and the PIPS projects. The dynamic process of collective action by rival firms creates the functional equivalent of demand articulation in a single firm. In other words, it can be conceptualized as 'collective articulation of demand'.

The demand articulation promoted by collective research should be viewed in and can be explained by the overall framework of industrial technological linkages. It can assist in creating a national technological infrastructure. Sometimes it results in the establishment of upstream technological linkages, and at other times it contributes to establishing downstream linkages. In other words, the advantages of collective research among rival firms should be viewed in a broad context and a long-term perspective. Viewed in this manner, one can see how collective action in research among rival firms leads to a positive-sum game.

The implications of these findings are not only limited to collective research. They might have much broader implications in terms of management principles for individual science and/or technology projects, or provide guiding principles for overall science and/or technology policies.

10. Concluding remarks

This chapter showed how demand articulation worked at the company level, industrial sector and country levels. The question which remains is whether demand articulation works at an international level. Examples of those problems on the international level are global environmental problems such as global warming. Within national boundaries, the question remains as to whether demand articulation can be applied to the public sector. Examples are social and urban development.

Therefore, I would like to conclude this chapter by presenting research agendas for future research. The first agenda is whether demand articulation is necessary and/or effective in technology development in international and public areas. The structure of demand articulation in these areas might be quite different from those in the areas described in this chapter. Who is the beneficiary of the development of those technologies? Who pays for the cost of their development? Is there any mismatch between those two agents?

The second agenda is to determine if the process of demand articulation is different, when it is applied on the international and public levels. Even if demand articulation is effective for such technology development, the decision-making process for adopting new technologies is different. For social and urban development, actors in decision making include citizens, developers and local government. For global environmental problems, national governments, regional organizations and international organizations are the decision-makers.

The third agenda is to determine who the demand articulation agent is. It might be local government and international organizations in urban development and global environment respectively. Those agents, however, do not own the technical capability for demand articulation. What is important is how they can mobilize and organize the technical capability needed for demand articulation.

There might be other research agendas. But my belief is that all of these are challenging to science policy researchers. To answer these challenges, the research community of science policy research might have to be extended to include other disciplines. In fact, how to best organize these needed disciplines is a challenging subject.

References

[1] R. Rowthwell et al., 'SAPPHO updated -SAPHHO phase II', *Research Policy*, vol. 3, no. 3, 1974.

[2] E. Hippel, *The Sources of Innovation*, Oxford University Press, 1988.

[3] N. Rosenberg, *Perspectives on Technology*, pp. 108–25, Cambridge University Press, 1976.

[4] R. Nelson and S. Winter, *An Evolutionary Theory of Economic Change*, pp. 254–62, The Belknap Press of Harvard University Press, 1982.

[5] S. Kline and N. Rosenberg, 'An overview of innovation', in *The Positive-Sum Strategy*, R. Landau and N. Rosenberg (Eds.), National Academy Press, 1986, pp.275–305.

[6] OECD, *Case Study of Electronics with Particular Reference to the Semiconductor Industry*, Joint Working Party of the Committee for Scientific and Technological Policy and the Industry Committee on Technology and the Structural Adaptation of Industry, pp.133–63, November 1977.

[7] F. Kodama, 'Demand articulation: targeted technology development', in *Technik-Politik: Angeschichts der Um welt-Katastrophe,* Ed. H. Krupp, pp. 273–94, Physica-Verlag Heidelberg, 1990
[8] Japan Industrial Policy Research Institute, *A Study of Innovation Process,* (in Japanese), Sangyou-Kenkyushyo, July 1989, pp. 48–57.
[9] N. Sawazaki, M. Yagi, M. Iwasaki, G. Inada and T. Tamaoki, 'A new video-tape recording system', *Journal of the Society of Motion Picture and Television Engineers,* vol. 69, pp. 868– 71, 1960.
[10] K. Oshima and F. Kodama, 'Japanese experiences in collective industrial activity: an analysis of engineering research associations', in *Technical Cooperation and International Competitiveness,* H. Fusfeld and R. Nelson (eds.), Rensselaer Polytechnic Institute, 1988, pp. 93–103.
[11] Y. Tarui, 'The records of the VLSI joint laboratory' (in Japanese), *Shizen (Nature),* Tokyo, September 1981.
[12] Engineering Research Association of Pattern Information Processing System, *Demonstration Panels: Integrated system prototype of pattern information processing system,* September 1980.
[13] J. Sigurdson, *Industry and State Partnership in Japan: The very large scale integrated circuits project,* Discussion Paper no. 168 (Research Policy Institute, University of Lund, Sweden, 1986), pp. 86–93.
[14] F. Kodama, 'Direct and indirect channels for transforming scientific knowledge into technical innovations', in *Transforming Scientific Ideas into Innovations: Science policies in the United States and Japan,* B. Bartocha and S. Okamura, (eds.), Japan Society for the Promotion of Science, 1985, pp.198–204.
[15] L. Fisher, 'Need for high-tech consortia stressed', *The New York Times,* C1–C3, 12 January, 1989.
[16] H. Fusfeld and C. Haklish, 'Cooperate R&D for competitors', *Harvard Business Review,* vol.85, no.6, 1985.
[17] F. Kodama, 'Technological diversification of Japanese industry', *Science,* vol. 233, pp.291–6, 1986.
[18] MITI, *Summary of International Research Cooperation* (in Japanese), Agency of Industrial Science and Technology, June 1984.
[19] I. Shirai and F. Kodama, *Quantitative Analysis on Structure of Collective R&D by Private Corporations in Japan* (in Japanese), NISTEP Report no.5 (National Institute of Science and Technology Policy, Tokyo, August 1989).
[20] M. Kobayashi, *A Mathematical Model of Collective Research,* MA Dissertation submitted to Graduate School of Policy Science, Saitama University, 1987.
[21] F. Kodama, 'Rivals' participating in collective research: its eonomic and technological rationale', in *Science and Technology Policy Research: What should be done? What can be done?,* pp. 141–64, The Proceeding of the international conference on science and technology policy research, H. Insoe, M. Kawasaki and F. Kodama, (eds), Mita Press, Tokyo, 1991.
[22] G. Loury, 'Market structure and innovation', *Quarterly Journal of Economics,* vol. 93, 1979.
[23] I. Lee and L. Wilde, 'Market structure and innovation: a reformulation', *Quarterly Journal of Economics,* vol. 94, 1980.
[24] P. Dasgupta and J. Stiglitz, 'Industrial structure and the nature of innovative activity', *Economic Journal,* vol. 90, 1980.
[25] A. Spence, 'Cost reduction, competition and industrial performance', *Econometrica,* vol. 52, pp. 101–21, 1984.
[26] M. Katz, *An Analysis of Cooperative Research and Development,* Discussion Paper (Princeton University, 1986).

Chapter 5
Innovation pattern: from breakthrough to technology fusion

1. Introduction

Japan has certainly been successful in designing products and production systems based on foreign basic technologies. However, it is often said that Japan is still an adapter as much as an innovator. [1] The study made by Gellman Associates [2] for the US National Science Foundation is often cited as proof for this view. The study found that Japanese scientists developed only 2 of 100 radical breakthroughs introduced into the marketplace between 1953 and 1973.

This study tried to compare the relative 'innovativeness' of the United States, Great Britain, West Germany, France, Japan and Canada. In order to identify cases of innovation, an international panel of experts was organized. They ranked each of 1,310 innovations in terms of importance, and identified the top 500 innovations. The distribution of the 500 innovations ultimately selected by the panel was as follows: the United States, 63 percent; the United Kingdom, 17 percent; West Germany, 7 percent; Japan 7 percent, France, 4 percent; and Canada, 2 percent.

Short histories were obtained for 380 of the original 500 innovations. In order to measure them on a scale of relative values, three principal categories of innovation were selected: radical breakthroughs, major technological shifts, and improvements. Of the 100 radical breakthroughs, the United States produced 65; the United Kingdom 25; France 4; and West Germany 3, but Japan produced only 2. However, of the 121 improvements, the US produced 98; West Germany 8; France 2 and the United Kingdom 2; but Japan produced as many as 10 innovations in this category.

However, almost twenty years have passed since then. Therefore, the argument that the Japanese are not innovators might be based on outdated evidence. We should ask what has happened since 1973. Were there increases in innovations by the Japanese during the 1970s and 1980s?

The evidence described above is based on a subjective selection of innovations and a subjective method of evaluating them, although the authors tried their best to make the study as objective as possible. A more objective methodology is needed to investigate the truth of the widespread perception that Japan is still a copycat country incapable of producing original work.

In order to determine if the copycat characterization of Japanese technology is valid today, F. Narin [3] analyzed the database of all US patents issued between 1975 and 1985. He concludes that the US patent statistics do not support the view that the Japanese are unoriginal copycats.

In terms of the size of the Japanese innovation activities, measured by patent counting in the database, he finds:

1. The large and increasing number of patents being issued to Japanese inventors demonstrates a burgeoning Japanese inventive vitality: the share held by Japanese inventors doubled from 8.9 percent to 17.9 percent of the total; the top three U.S. patent recipients were Japanese manufacturers; and, Japanese inventors obtained more patents in the United States than inventors in the United Kingdom, France, and West Germany combined.
2. The counting of the Japanese patents by USPO (US Patent and Trademark Office) classification, reveals that areas of Japanese technological strength are concentrated on high technologies: the Japanese held 57.6 percent of the patents in the photography classification, 43.6 percent of photocopying patents. In other areas that are less actively patenting but none the less technologically crucial, the results are the same: they held 40 percent of the patents in the dynamic magnetic information storage and retrieval, 43.2 percent in dynamic information storage and retrieval.

In terms of the quality of Japanese innovation activities, the examiner citations in US patents granted from 1975 through 1984 that were made to 1975 through 1982 US patents, reveals:

3. It is clear that there is a higher proportion of Japanese-held patents in the listings of highly cited patents than we would expect statistically: 13.9 percent of all 1982 patents fall into the top 10 percent highly cited category, and 19.1 percent of Japanese-held patents fall into this category; the ratio of these two values is 1.37, indicating 37 percent more Japanese patents in the highly cited list than statistically expected; and the corresponding figures are 1.06 for American-invented patents, 0.94 for UK-invented patents, 0.84 for French, and 0.79 for West Germany.
4. A count of examiner citations to the non-patent literature, most of which can be assumed to be scientific works such as books, monographs and articles, suggests that although the Japanese may be highly inventive, their inventions are not based on scientific research as much as those associated with Western technological efforts.

Comparing these two studies, i.e. Gellman's and Narin's, we observe a tremendous gap between perception and revealed reality. The gap is so large that it is hard to believe it originates solely in differences in methodology and/or differences in the time periods observed. There may be a more profound reason.

Between the two time periods, there might have been a drastic change in patterns of innovations. Before going into detail, we need a deeper understanding of the intrinsic nature of technology. In his chapter: 'Technological interdependence in the American economy' in his book *'Inside Black Box: Technology and economics'*, [4] Nathan Rosenberg wrote:

> The public image of technology has been decisively shaped by popular writers who have been mesmerized by the dramatic story of a small number of major inventions – steam engines, cotton grins, railroads, automobiles, penicillin, radios, computers, and so on. In addition, in the telling of the story, overwhelming emphasis is placed on the specific sequence of events leading up to the decisive actions of a single individual. Indeed, not only our patent law but also our history textbooks and even our language all conspire in insuring that a single name and date are attached to each invention.

The missing link, he argues, is the lack of attention given to the network of larger technological relationships in which specific inventions are always embedded. He then formulated significant characteristics of technologies: complementarities, cumulative impacts and interindustry relationships.

He describes *complemetarities*: invention hardly ever functions in isolation; it depends on the availability of complementary technologies; the productivity growth of industrial economies is the complex outcome of large numbers of interlocking, mutually reinforcing technologies; the relevant unit of observation is seldom a single innovation but an interrelated clustering of innovations; even apparently spectacular breakthroughs usually have only a gradually rising productivity curve; and the combined effects of large numbers of improvements within a technological system may be immense.

He argues the *cumulative impact of small improvements*: a large portion of the productivity growth takes the form of a slow accretion of individually small improvements in innovations; and much of the technological change that goes on in an advanced industrial economy is, if not invisible, at least of a low-visibility sort.

He describes *interindustry relationships*: the benefits of increased productivity flowing from an innovation are captured in industries other than the one in which the innovation was made; a full accounting of the benefits of innovation must include an examination of interindustry relationships; industrial development under a dynamic technology leads to wholly new patterns of specialization both by firm and by industry; and it is impossible to compartmentalize the consequences of technological innovation even within conventional industrial boundaries.

Rosenberg summarizes by saying that the inability to take these characteristics of technologies fully into account is a fundamental limitation of most of the recent literature on technological innovation. In fact, technological progress in one sector of the economy has become increasingly dependent upon technological change in other sectors. Such interindustry flows of technology are one of the most distinctive characteristics of advanced industrial societies.

In this chapter, I would like to ask whether Rosenberg's characterization of modern technologies is sufficient to explain the gap which can be found between Gellman's and Narin's Studies.*First*, the case studies on which Rosenberg's characterization is based do not include recent civilian high technologies in which Japan has demonstrated her international competitiveness. *Second*, the difference in the observation time periods between Gellman's study and Narin's, i.e. between 1953 and 1973 and 1975 and 1985, is more significant than the difference in observation periods would indicate, because a techno-paradigm shift from modern technologies to high technologies occurred in the mid-1970s.

Japanese experiences might be creating a new type of innovation pattern. In the end of this chapter, I will also try to integrate several findings from the preceding chapters into the shift in innovation pattern.

2. Mechatronics revolution

In 1975, the Japanese created a new word, 'mechatronics', by combining the two words mechanics and electronics. Essentially it implies the following two categories of products:

- The marriage of electronic technology to mechanical technology resulted in the birth of a more sophisticated range of technological products. Typical examples are NC (numerically controlled) machine tools and industrial robots.
- Products in which a part, or the whole, of a standard mechanical product was superseded by electronics. Typical examples are digital clocks and electronic calculators.

In order to measure the diffusion of mechatronics technology in the various types of machinery, an indicator is needed. In the machine tool case, the diffusion rate can be measured by the ratio of the numerically controlled machine tools to the total production of machine tools. A marked increase in the diffusion rate, in fact, occurred in 1975. Since then, diffusion was quite rapid in this industry.

By the same token, a 'mechatronized machine' is defined as a machine with computer control. The diffusion rate can be measured by the ratio of the mechatronized machines to the total production of the machines. [5] On the basis of this measure, the diffusion curves of mechatronics in various categories of machines are obtained as shown in Figure 5.1. Those categories of machinery in which the diffusion rate of mechatronics is above 30 percent are industrial robots, machine tools, bending machines, and printing and bookbinding machinery. Those categories of machines which are not yet widely mechatronized are woodworking machinery, plastics-processing machinery, packaging machines and food-processing machines.

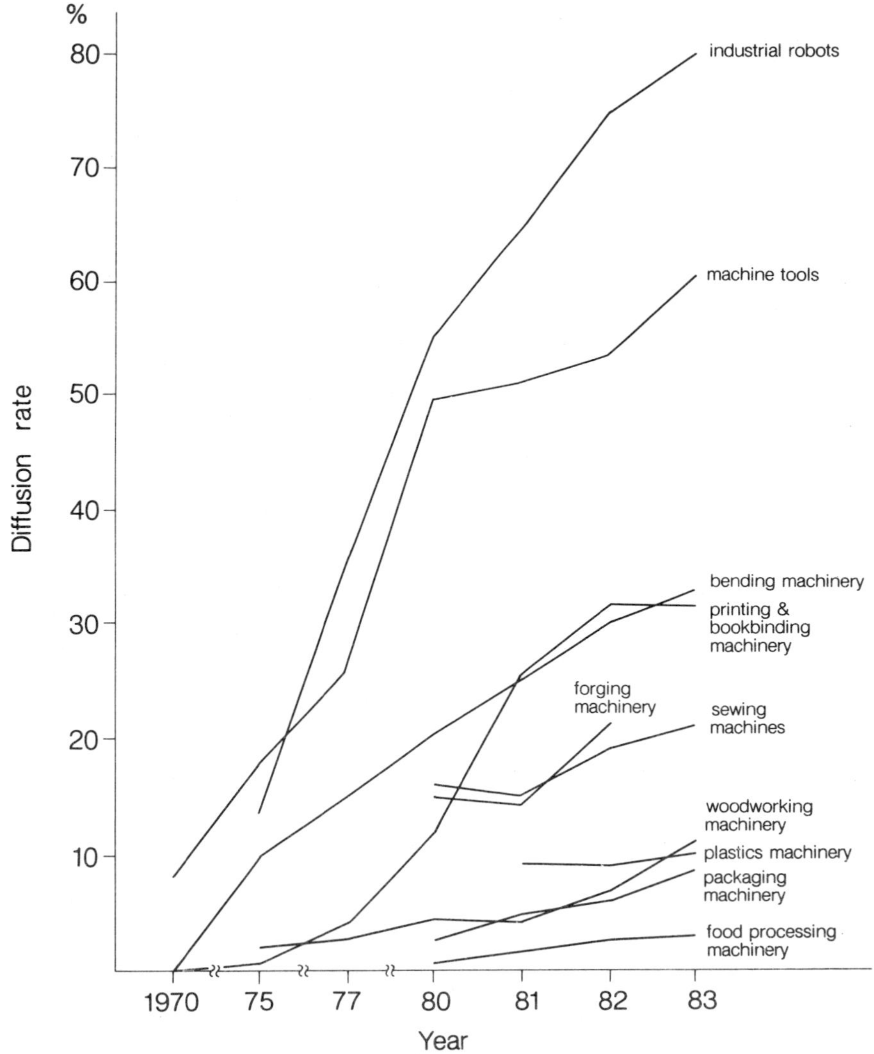

Figure 5.1 Diffusion curves of mechatronics technology

We also observe a significant difference in the growth rate of production between these two categories of machinery. The group with a diffusion rate of mechatronics higher than 30 percent has a higher growth rate. On the other hand, the group with diffusion of less than 30 percent is lower in growth rate. Thus there seems to be a positive correlation between the diffusion rate and the growth rate. This indicates the possibility that the group whose growth is stagnant can regain a growth momentum with the introduction of mechatronics technology.

The NC (numerically controlled) revolution in the Japanese machine tool industry has become possible only with cooperation from three other industries outside the machine tool industry itself. The Japanese are responsible for two inventions crucial to the machine tool industry: an improved servo motor which converts a small mechanical force into a large one; and a numerical controller which is compact, simple to use and cheap to make. [6]

The roots of machine tools with unprecedented levels of precision and reliability, go back thirty years. MIT (Massachusetts Institute of Technology) invented the numerical controller in 1952. This device had 2,000 valves. It controlled a machine tool automatically, but it was both huge and expensive, putting it out of reach of all but the biggest industries such as aircraft manufacturers.

Fanuc, a spin-off from Fujitsu, a supplier of communication equipment, designed a new type of servo motor which was called an electro-hydraulic stepping motor, in which electrical control pulses are converted directly into mechanical movements known as steps. It removed complications, and in particular it eliminated the need for a feedback loop.

The numerical controller developed by MIT needed a big computer capable of sophisticated calculations to transform complex curves into numerical values. Therefore, the MIT controller was the size of a small room. However, Fanuc made it not much bigger than a chest of drawers.

Fanuc's controller bases its operation principle on the scientific fact that it is possible to reduce most technical drawings to arcs (which can be expressed as a radius with starting and ending points) and straight lines (which can be defined as two points). On this basis, Fanuc developed a much simpler machine, which translates arcs and lines into pulses, than the one from MIT.

However, the device which harnesses the stepping motor to the worktable in a NC machine tool is called a ballscrew, which was developed by NSK(Nippon Seiko), Japan's leading maker of bearings. Its predecessor, the friction screw, wears with use. Wear introduces error, which gradually gets bigger and backlash appears, ultimately making the machine impossible to control. The great advantage of the ballscrew is that because lubricated ball bearings are inserted between the screw's nut and bolt its friction is very low. This means that the screw's mechanical characteristics do not change over the course of its life. There is little point in computers providing perfect reproducibility if the character of the parts of a machine tool alters in the middle of a production run. General Motors was the first to apply ballscrews during the 1940s in the steering gear of army trucks. Just at the time when NSK started making ballscrews after General Motors' patent expired, Fanuc asked it to make a ballscrew with perfect pitch.

In conclusion, without the development of a ballscrew with perfect pitch, it would never have been possible to hook up Fanuc's servo motor in an open-loop servo system.

A further contribution came from the material suppliers. It was the coating of 'Teflon' on the sliding bed of the machine tool, thus enabling the hooking up of the servo motor, which is good for precise adjustment but weak in torque. This also made low speed but uniform movement possible, an absolute necessity for the operation of a machine tool.

In 1972, Fujitsu decided to concentrate on making computers. It split Fanuc off as an independent entity, with Fujitsu holding 35 percent of the new company's shares. By that time, realizing that numerical control gave their products a competitive edge, the makers of machine tools had begun to buy Fanuc's controllers. But it was clear to Fanuc that the market in big companies was too small; if Fanuc wanted to grow, it had to provide controllers that small and medium-sized companies with fewer than 300 employees could afford. It increased production to reduce the price.

In 1986, Funac produced 5,000 controllers a month – more than half the world's production – and 20,000 servo motors. No other maker produced controllers in such numbers. Production in volume gave Fanuc profit margins of 40 percent before tax and made it one of the most profitable companies in Japan. The number of controllers Fanuc produced also made it economically feasible for the company to produce custom-made chips, which in turn enabled controllers to become smaller. Thus Fanuc's machines have all their electronics on a single-printed- circuit board.

3. Optoelectronics revolution

Optoelectronics, a marriage of electronics and optics, is already yielding important commercial products such as optical fiber communication systems. It united the electron with the ephemeral photon, the particle of light, to attain greater efficiency in data processing and transmission than electronics can achieve by itself. It is drastically revolutionizing the communication system and widely expected to form the next generation of information-based technology.

In 1986, *Fortune* magazine [7] asked ten scholars, business executives, government officials and foundation leaders in each field to rank the state of research and development in the UnitedStates, Japan, Western Europe and the USSR on a scale of 1 to 10. It studied the following four technological fields: computers, chips, and factory automation; life sciences; advanced materials; and, optoelectronics, as shown in Table 5.1.

As can be seen in the Table 5.1, there is one exception in which the United States was not rated as no. 1: it was optoelectronics. The magazine reported that everyone conceded that the Japanese lead the world hands-down in this important new technology which was originally developed

Table 5.1 Results of *Fortune* scoreboard

Technical field	USA	Japan	W Europe	USSR
computers, chips, factory automation	9.9	7.3	4.4	1.5
life sciences	8.9	5.7	4.9	1.3
advanced materials	7.7	6.3	6.0	3.8
optoelectronics	7.8	9.5	5.7	3.6

Source: Gene Bylinsky, 'The high tech race: who's ahead?' *Fortune*, 13 October, 1986, pp. 18–36.

in the United States It also reported an expert's comment, saying 'the Japanese lead is very considerable. And there's little evidence that anything we are doing in this country will close the gap in the near future'.

In his work on technological changes in the machine tool industry from 1840 to 1910, which he tried to approach from the purely technological level, Rosenberg finds that complex technologies create internal compulsions and pressures which, in turn, initiate exploratory activity in particular directions.[8] He states:

> A concept of technological disequilibrium may be helpful here. At any time, the component parts of a machine vary in their ability to exceed their present level of performance, which is determined by the capacity of some limiting component. Any important improvement in the operation of a component, whether it be the currently limiting one or not, is likely to create new obstacles, in the form of limitations imposed by another component, to the achievement of a higher level of performance. Thus single improvements tend to create their own future problems, which compel further modification and revision.

In describing the recent Japanese development in optical fibers, therefore, I will try to structure this mini-history around the concept of technological disequilibrium. However, at the same time, I will also try to show that this technological revolution became possible by a concerted effort of the different industries involved. [9]

An optical communication system consists of three technological components: a light-emitter (electrical to optical convertor), an optical fiber (transmission media), and a light receiver (optical to electrical convertor). Therefore, for the realization of optical communication systems, innovations in laser and semiconductor technologies had to be matched with those of optical fibers. In fact, the technology development in those two areas progressed along with that of optical fibers in an interactive and complementary manner as shown in Figure 5.2.

The first optical fiber had attained minimum transmission loss at the short wavelength of 0.8 μm. Therefore, around the later half of 1970s, a highly reliable semiconductor laser (GaAlAs laser) which emits light at this wavelength, was developed. And an optical communication system

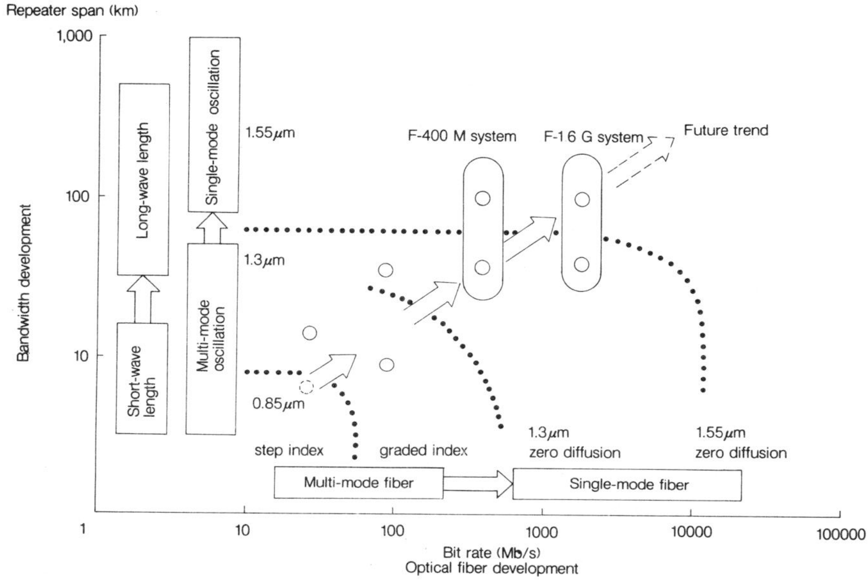

Figure 5.2 Technological progress in optical communication systems

with SiAPD as the receiver, was developed for a feasibility test of optical communication.

A new optical fiber manufacturing technology, a more efficient VAD was developed, and replaced the older MCVD. Key to this new technology was the complete elimination of OH content. This was done by developing a perfect dehydration sintering technology. As a result, the wavelength of minimum transmission loss shifted from a short wavelength to a long one of 1.3 μm $\approx$ 1.55 μm. Thus, a laser of long wavelength (GaInAs laser) and receivers of Ge-APD and InGaAs-APD were developed. All of these contributed to the construction of the Japanese optical fiber communication line, which runs the length of the Japanese island.

Thereafter, optical fiber of ultra small transmission loss was developed for an undersea cable line to be used for international communication. For the emitter, a single-mode laser (single spectrum) such as DFB-type and DBRO-type was developed. As a result, long distance and large capacity communication with the minimum transmission loss at 1.55 μm has become possible.

As can be seen in the above example, technological advances have not been necessarily attained in a simultaneous and parallel way in the three component areas, i.e. emitter, transmission and receiver. Instead, innovation in one area set the technological development agendas in the other two areas. In other words, a series of dynamic relations of demand articulation exists among three component technologies.

Moreover, as described in Chapter 3, the Japanese development of optical communication technology was conducted in a series of joint research efforts between different companies in different industrial sectors. Among others, those companies actively involved in joint efforts include NSG, NEC, SEI, and NTT. In fact, the three companies, NSG, NEC, and SEI, belong to the same industrial group, the Sumitomo Group, which may have provided a useful forum for integrating individual technology development efforts. And NTT is of course a most important customer for these three companies.

4. Technology fusion

4.1 Modern technology versus high technology

As described before, through his extensive studies of the history of technologies, N. Rosenberg finds that complementarities, cumulative impacts and interindustry relationships are significant characteristics of modern technologies. [4]

However, as was clearly alluded to in the two case studies in mechatronics and optoelectronics, we need an enhanced version of this concept to characterize high technologies. Therefore, I would argue that these three characteristics belong to high technologies in embryo, because high technologies as such were not obviously included in Rosenberg's studies. Although the characteristics of high technologies might embody these same characteristics, the way in which technologies are produced has changed.

Therefore, to describe high-technology fully we need to develop a new theory which, while encompassing Rosenberg's insights, also incorporates the newly emerging characteristics of high technologies. These characteristics differ from those which Rosenberg identified in modern technology, in the following ways.

First, Rosenberg's characteristics were derived from long-term historical studies, in which these phenomena were discovered as historical coincidences, i.e. unintended occurrences and results of passive actions. However, mechatronics and optoelectronics were developed fairly quickly, in a series of active joint research projects among different companies in different industrial sectors. This indicates that the institutional framework for innovation has changed.

Second, the three characteristics are derived from retrospective studies, and thus were not recognized clearly during the technology development, even in the successful cases. However, as our case studies depict, those characteristics were recognized, even in the early stage of the development of mechatronics and optoelectronics, and appropriate actions were taken to accommodate these inevitable characteristics of the technologies. This indicates that the level of understanding innovation dynamics has been enhanced.

Third, the diffusion of mechatronics technologies was surprisingly rapid throughout the Japanese manufacturing economy, as shown in Figure 5.1 of the preceding section. In fact, in less than fifteen years diffusion had reached saturation. And diffusion was almost contemporary among different manufacturing industries. On the other hand, based on his study of technical change in the machine tool industry from 1840 to 1910, Rosenberg suggested that the machine tool industry may be regarded as a center for the acquisition and diffusion of new skills and techniques in a machinofacture type of economy, as a result of technological convergence. [10] And its diffusion occurred in a sequential manner from firearms, sewing machines, bicycles, through automobiles. This suggests that diffusion framework has also changed.

Therefore, we might better say that: the mechatronics revolution is generated by the fusing of mechanical technology with electronic and material technologies, and the optoelectronics revolution is generated by the fusing of glass technology with cable and electronic device technologies. As the names mechatronics and optoelectronics implies, 'fusion' means a more than the summation and combination of different technologies, and it implements an arithmetic in which one plus one makes three.

The fusion of technologies goes beyond mere combination. Fusion is more than complementarities, because it creates a new market and new growth opportunities for each participant in the innovation. Fusion goes beyond the cumulation of small improvements, because it blends incremental improvements from several (often previously separate) fields to create a product endowed with some extra ingredient not found elsewhere in the market. It also goes beyond interindustry relationships, because different innovations in different industries progressed in parallel with each other, taking the form of joint research.

4.2 Technical breakthrough versus technology fusion

Contrasting characteristics of high technology with those Rosenberg identified in modern technology helps clarify the innovation mechanisms associated with technology fusion. A second contrast between fusion and technical breakthroughs will help clarify the institutional effects and facilitators of fusion-type innovation.

On the basis of the experiences described above, I would argue that there are two fundamental types of innovation: one is the 'technical breakthrough' and the other is a 'technology fusion'. [11]–[13] These are exemplified by the transistor revolution, and the mechatronics revolution, respectively.

Let us now characterize technological fusion *vis-à-vis* technical breakthrough. *First*, breakthrough innovations are associated with strong leadership in a particular industry, and fusion innovations become possible through concerted effort by several different industries. Developments in the Japanese machine tool industry are a case in point. As described above, the mechatronics revolution in this industry became possible only with the cooperation of three other industries outside machine tools itself.

Secondly, fusion-type innovation contributes not to the radical growth of certain companies, but to the gradual growth of all the companies in the relevant industries. The semiconductor revolution of the late 1950s which occurred in the United States, brought about a major reshuffle in the relevant industries. Many of the existing vacuum-tube manufacturers went out of business, while some of the new entrants such as Texas

Table 5.2 US semiconductor market shares of the major firms, selected years, 1957–66

Type and name of firm	Percentage of market			
	1957	1960	1963	1966
Western Electric	5	5	7	9
Receiving tube firms				
General Electric	9	8	8	8
RCA	6	7	5	7
Raytheon	5	4	b	b
Sylvania	4	3	b	b
Philco-Ford	3	6	4	3
Westinghouse	2	6	4	5
Others	2	1	4	3
subtotal	31	35	25	26
New firms				
Texas Instruments	20	20	18	17
Transitron	12	9	3	3
Hughes	11	5	b	b
Motorola	b	5	10	12
Fairchild	b	5	9	13
Thomson Ramo Wooldridge	b	b	4	b
General Instrument	b	b	b	4
Delco Radio	b	b	b	4
Others	21	16	24	12
subtotal	64	60	68	65
total	100	100	100	100

Source: J. E. Tilton, *International Diffusion of Technology*, The Brookings Institution, 1971.

Note: b = Not one of the top semiconductor firms for this year. Its market share is included under 'others'.

Instruments, Motorola and Fairchild, grew rapidly. In 1966, the market share of the new entrants was 65 percent, while that of the previous vacuum-tube manufacturers was only 26 percent, [14] as shown in Table 5.2. This is a vivid manifestation of a 'creative gale of destruction' la Schumpeter.

On the other hand, in the ten years from 1975 to 1985 Japanese production of machine tools rose from fourth in the world after the USSR to first, as shown in Figure 5.3. This leapfrog growth in the world market and in such a short time period would never have been possible without a substantial innovation which had been brought into the industry. However, there was no substantial change of market share among existing Japanese machine tool firms during this period. In other words, Japanese machine tool manufacturers all grew in this decade.

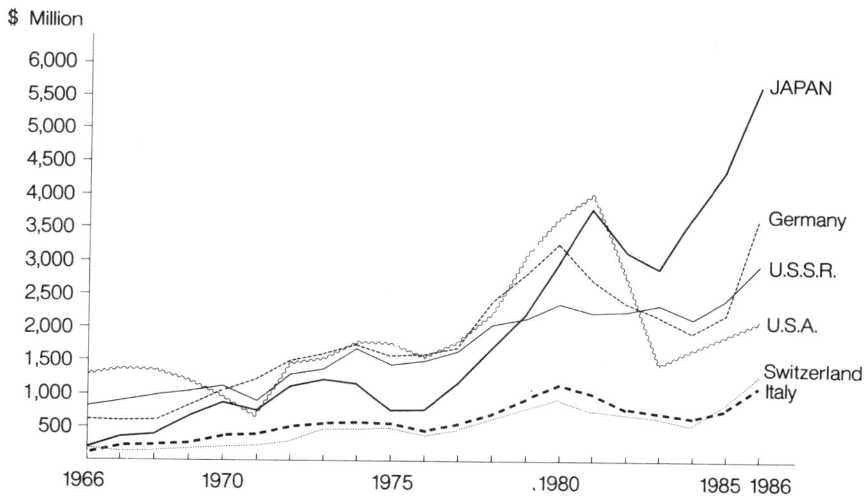

Figure 5.3 Production of machine tools in major countries *Source*: *American Machinist*.

Third, the breakthrough-type innovation is associated with defense policy, as exemplified by the US development of microelectronics and civilian airplanes. Although several authors expressed skepticism about the importance of relationships with the defense sector in cases such as the transistor [15] and integrated circuits, [16] even in aircraft, [17] at the very least we can say that this relation is subtle and indirect. On the other hand, it is obvious that fusion-type innovation can be induced by industrial policy. Let us look into the development of the industrial policy which induced the mechatronics revolution of the mid-1970s.

Before 1971, the Law on Temporary Measures for the Development of

Machinery Industry (enacted in 1956) and the Law on Temporary Measures for the Development of Electronics Industry (enacted in 1957) were enforced independently of each other. The main content of these laws was low-interest loans given by semi-governmental financial institutions, such as the Japan Development Bank, to companies for technology development projects, particularly for commercialization projects of new technologies.

In 1971, however, these two laws were transformed into the Law on Temporary Measures for the Development of Specified Machinery and Electronics Industries, as shown in Table 5.3.

Table 5.3 Chronology of legislation of mechatronics-related laws

1957		Law for machinery industry	Law for electronics industry
1971		Law for the development of specified machinery and electronics industries	
1975	Mechatronics revolution		
1978		Law for the development of specified machinery and information industries	
1985		Expired	

The mechatronics revolution was clearly envisioned in this law, because expressions such as 'consolidation of machinery and electronics into one' or 'systematization of them', were used by MITI for explaining the objectives of the law in the Diet. [18] In the Diet, the Minister of MITI explained the objectives of the law as follows:

In deciding upon an Intensification Plan, the competent minister shall take note of the increased inter relationships among different industries, in particular, between the machinery and the electronic machinery and equipment, and shall pay due consideration to the direcion of so-called 'consolidation of machinery and electronics into one' or 'systematization of them'.

In 1978, this law was transformed into the Law on Extraordinary Measures for the Development of Specified Machinery and Information

Industries, which expired in 1985. Although there are doubts about how effective these laws were in influencing industry behavior, the merger of the two laws had at least an indicative meaning to the industries involved, in the sense that they showed clearly where major innovations were to occur in the 1970s

5. Mathematical formulation

The process of technology fusion begins with industry's interests in product fields other than its principal ones. Therefore, the traditional dichotomy between product and process innovation is not helpful in analyzing technology fusion. A more relevant dichotomy is whether or not a sector's industrial R&D is done within or outside its principal product fields, because industrial R&D activity outside principal product fields is directed toward creating technology fusion.

Again, the rich Japanese R&D data source in the *Survey of Research and Development* by the Statistics Bureau of the Prime Minister's Office, makes it possible for us to operationalize the above described conceptualization into an empirical study because companies' intramural expenditure on R&D is disaggregated into thirty-one different product fields. [19]

5.1 Reciprocity

Simply stated, the mechatronics revolution was realized as a result of the fusion between mechanical technology and electronics technology. Thus the minimum unit of technology fusion is the pair of two different industrial sectors, say, *i-th* industry and *j-th* industry. Therefore, the unit of analysis is a sector's R&D investment into another industry's principal product fields, say, *i-th* industry's R&D investment into *j-th* industry's principal product fields.

However, one-way investment might be nothing but a diversification effort and may not lead to innovation. [20] On the other hand, if the investment is done reciprocally, it might lead to the creation of a new technological area, and possibly to innovation. For example, the ceramics industry had invested in product fields such as ordinary machinery, electrical machinery and electronics. However, 'new ceramics' was created only after the machinery and electronics industries began to invest in ceramics.

It could be argued that this hypothesis of reciprocal research investment is more than simply an operational definition for technology fusion. In other words, if cross investment is simply the mixing together of different types of technology, it would be possible simply through the merger of corporations possessing different types of technology. Such a combination of different technologies through mergers, however, would merely involve one technology making use of another, with little likelihood of bringing about the technology fusion discussed here. On the contrary, if reciprocal investment takes the form of two corporations undertaking joint research, the two technologies would be of equal importance and technology fusion could easily result. The hypothesis of reciprocity is thus intrinsic in technology fusion itself.

Therefore, I would argue that the essence of technology fusion is its reciprocity. However, the requirement for reciprocal investment might not be a sufficient condition for technology fusion, but it is obviously a necessary condition. On this basis, we can assume that technology fusion between *i-th* industry and *j-th* industry is realized only after the investment is done in two ways, i.e. *i-th* industry's investment into *j-th* industry's principal product fields and *j-th* industry's investment into *i-th* industry's principal product fields.

5.2 Substantiality

Another requirement for technology fusion to be realized, is that a sector's R&D investment into another industry's principal fields should be substantial.

Considering the dynamics of R&D programs described in Chapter 3, we can assume that the research reaches the development stage only if the expense exceeds a certain amount. What is important here for this purpose is the demarcation between the exploratory phase and the development phase.

In Chapter 3, we built a survival model of an R&D program's investment as follows:

Let

$R(C)$ = the probability that an R&D program can survive until its annual investment reaches the amount of C,

$r(C)$ = the canceling probability of an R&D program whose investment level is C,

then $r(C)$ can be formulated as the probability that an R&D program is canceled, given that it can survive up to the investment level of C, therefore, the probability $r(C)$ can be represented by,

$$r(C) = f(C)/R(C) = -R'(C)/R(C).$$

And, in Chapter 3, by assuming that the canceling probability function $r(C)$ is a decreasing function of C, for each industrial sector, we identified the following three types of function:

$r(C) = r$, for science-based industry;

$r(C) = a \cdot \exp(-b \cdot C)$, for conventional industry; and,

$r(C) = a.\exp(-b \cdot C) + g$, for high-tech industry.

The demarcation can be done on the basis of the canceling probability. We can define an R&D program in the development phase as one whose canceling probability is smaller than a specified value r^*. Then, based on each industry's estimated relation between the canceling probability (r) and the annual investment (C), as identified in Chapter 3, we can identify the boundary value of annual investment, beyond which an R&D program can be supposed to enter the development phase, as shown schematically in Figure 5.4.

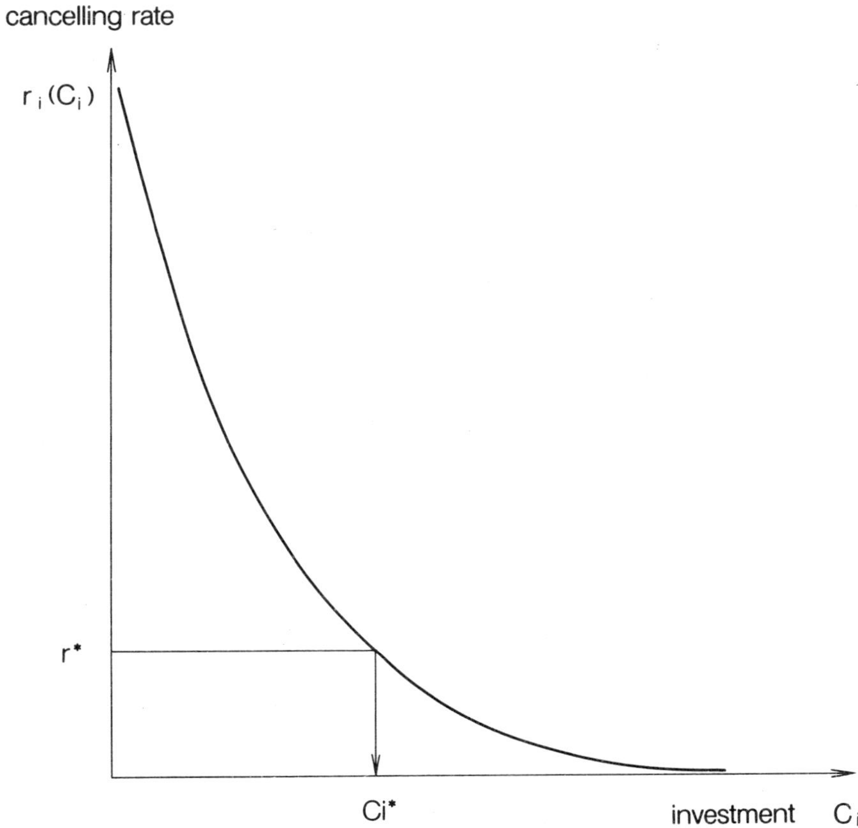

Figure 5.4 Calculation of threshold value for *i-th* industry

However, the boundary value for industrial chemicals cannot be obtained because this industry is classified as a science-based industry. Therefore, in the calculating the boundary value for it, we use the canceling rate function for conventional industry, although the level of significance is not satisfactory. Thus, for each industrial sector, we can obtain the threshold value C_i^*, which demarcates exploratory search from advanced development.

5.3 Technology fusion graph

Based on these two requirements for technology fusion to be realized, i.e., reciprocity and substantiality, we can formulate a model of the realization mechanism of technology fusion as follows:

Let

R_{ij} = *i-th* industry's R&D investment in the *j-th* industry's principal product fields,

then technology fusion is assumed to be *realized* between the *i-th* and *j-th* industry, only if

$R_{ij} > C_i^*,$

and

$R_{ji} > C_j^*$

Thus we could construct a model of technology fusion associated with a given pair of industries. Our interest is then directed to analyzing the network structure of technology fusions. The best way is to describe the structure of technology fusion graphically. In the non-directed graph, each industry is represented by a vertex, and an arc is drawn between a pair of two vertices only if there is *realized technology fusion* between the two industries represented by the corresponding two vertexes. We can call this graph a 'technology fusion graph'. And such a graph can be constructed for each year since 1970.

Therefore, we can formulate a technology fusion graph by a non-directed mathematical graph, $F = [F_{ijt}]$, where,

$$F_{ijt} \begin{cases} = 1 \text{ if } R_{ijt} > C_i^* \text{and } R_{jit} > C_j^*, \\ = 0 \text{ otherwise.} \end{cases}$$

5.4 Determination of threshold values

In order to draw technology fusion graphs, we have to determine the critical value of the canceling rate ($\overset{*}{r}$) which demarcates the development and exploratory phases. Since there is no information available, its determination can only be done on an empirical basis. In other words, it can be *calibrated* in the critical canceling rate.

It is said that the Japanese created a new word 'mechatronics' in 1975, by combining mechanics and electronics. If the development of mechatronics technology is measured by means of the ratio of the NC machine tools to the total production of machine tools, a marked increase did occur in 1975, as described before.

However, considering that the term mechatronics encompasses NC machine tools, industrial robots and digital clocks, we can identify the realization of the mechatronics revolution by the persistent existence of the quadruple connection among ordinary machinery, precision instruments, electrical machinery, and communication and electronic equipment, i.e. all the possible arcs between every pair of the two industries out of the quadruple.

Therefore, we can determine the critical canceling rate so that the quadruple connection begins to appear in 1975 and is maintained after that. In other words, before 1975 we have imperfect quadruples in which some arcs are missing, but after 1975 we have the complete quadruple. Through this kind of calibration, we found that the critical value of the canceling rate r^* should be set to be 0.07. We can then assume that the R&D program whose canceling rate is smaller than 0.07 is in the development stage, and on this basis we can calculate the threshold value of annual R&D investment for each industrial sector. The results of this calculation are shown in Table 5.4.

Table 5.4 Identification of threshold values

Industrial sector	Threshold value (¥ million)
food manufacturing	467
textile mill products	822
pulp and paper products	365
printing and publishing	626
industrial chemicals	1834
oil and paints	579
drugs and medicines	300
other chemical products	648
petroleum and coal products	369
rubber products	695
ceramics	984
iron and steel	639
non ferrous metals and products	1003
fabricated metal products	773
ordinary machinery	656
electrical machinery	231
communication and electronics	355
motor vehicles	948
other transport equipment	650
precision equipment	382

Once the critical value of the canceling rate is determined, we can draw a technology fusion graph for each year since 1970. Through all the years studied, technological fusions are clustered around three major industries – ordinary machinery, electrical machinery and industrial chemicals. The first two industries belong to the fabrication industry, while the last one belongs to the material industry. In 1970, three clusters are separated from each other, as shown in Figure 5.5. There is no technology fusion realized among these three clusters.

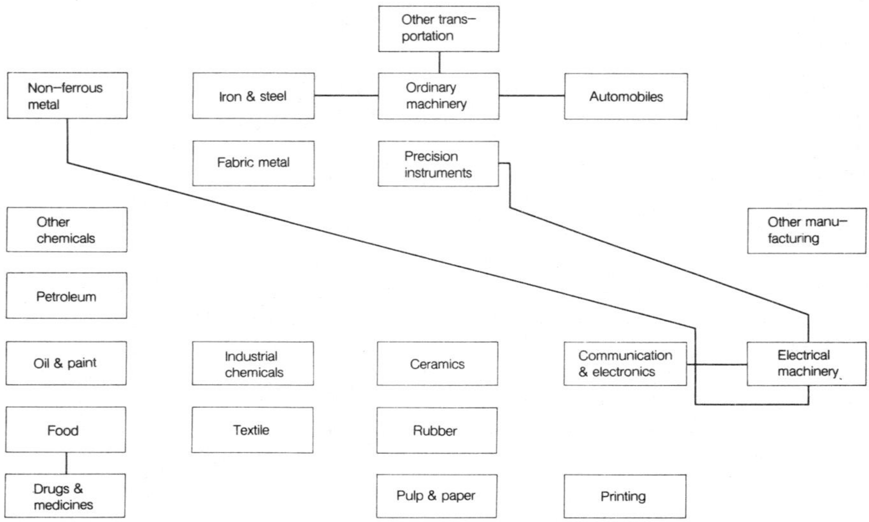

Figure 5.5 Technology fusion graph (in 1970)

6. High technology and technology fusion

On the basis of technology fusion graphs drawn as we defined, we found that almost all the so-called high technologies can be produced by technology fusion. Technology fusion graphs for 1974, 1975, and 1982 are shown in Figures 5.6 to 5.8. By using those graphs, it will be shown that most of the high technologies are the products of fusion-type innovation. Specifically, we will describe how such high technologies as biotechnology, mechatronics and new ceramics were developed.

In 1974 (Figure 5.6), a triangle connection emerged among food, drugs/medicines and industrial chemicals. This can be interpreted as the emergence of 'biotechnology'. The dual connection between food and drugs/medicines, which is supposed to indicate fermentation technology, had already appeared in 1970, as shown in Figure 5.5. However, fermentation technology alone lacked the components of a science-based industry. Thus, when industrial chemicals was added to this network, biotechnology was established as a high technology area. This technology fusion component of biotechnology might explain why it is said that Japan will become a front-runner in a much shorter time than expected.[21]

As far as 'mechatronics' is concerned (Figure 5.7), technology fusion between ordinary machinery and electrical machinery began in 1971.

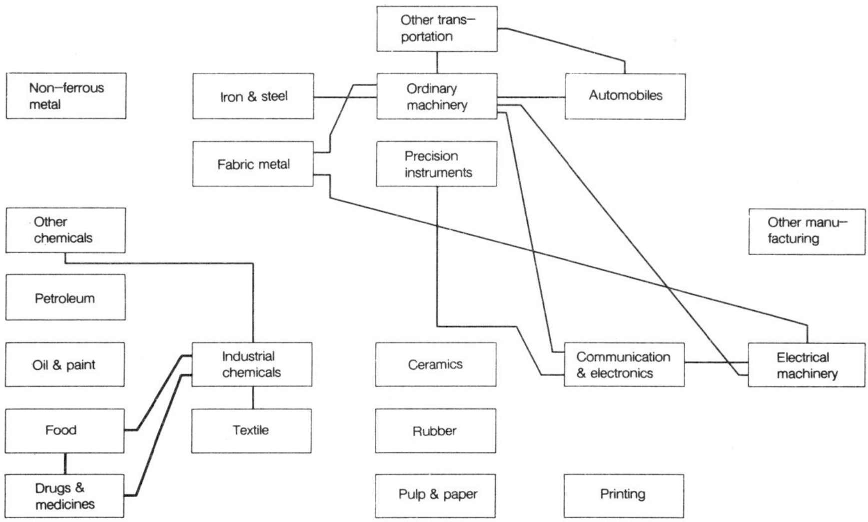

Figure 5.6 Technology fusion graph for biotechnology (in 1974)

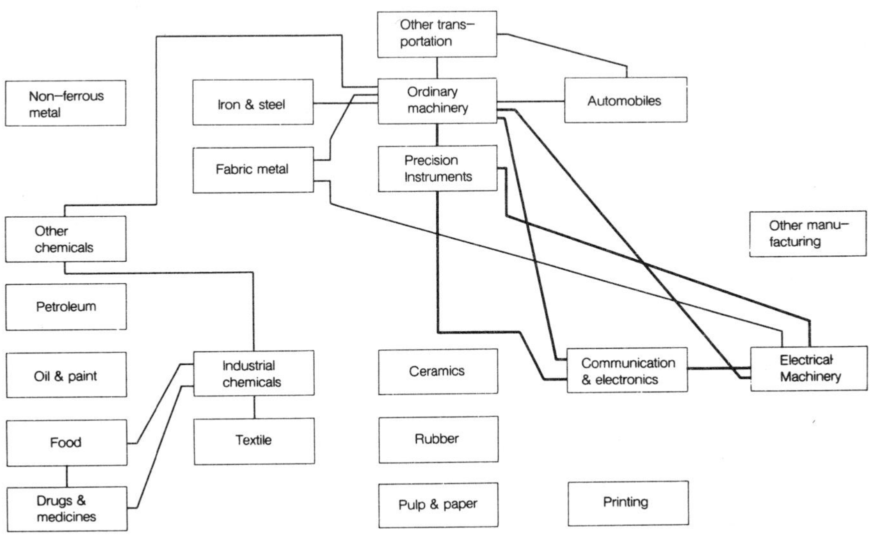

Figure 5.7 Technology fusion graph for mechatronics (in 1975)

Around 1973, communication and electronics joined this grouping and thus formed the triple connection among the three industries which indicates an early version of mechatronics. In order to perfect mechatronics technology, however, precision instruments must be included. And it was not until 1975 that the quadruple connection began to appear persistently among ordinary machinery, precision instruments, electrical machinery, and communication and electronic equipment.

In 1982 (Figure 5.8), technology fusion between ceramics and ordinary machinery and between ceramics and electrical machinery appeared. These connections represent the emergence of 'new ceramics'. The connection between ceramics and ordinary machinery, which represents 'fine ceramics', was formed in 1980. Subsequently, the connection between ceramics and industrial chemicals, which indicates the emergence of the new material, was formed in 1981. In 1982, the connection between ceramics and electrical machinery was added to the connections already established. Thus, a new materials revolution was realized in 1982, as far as ceramics is concerned.

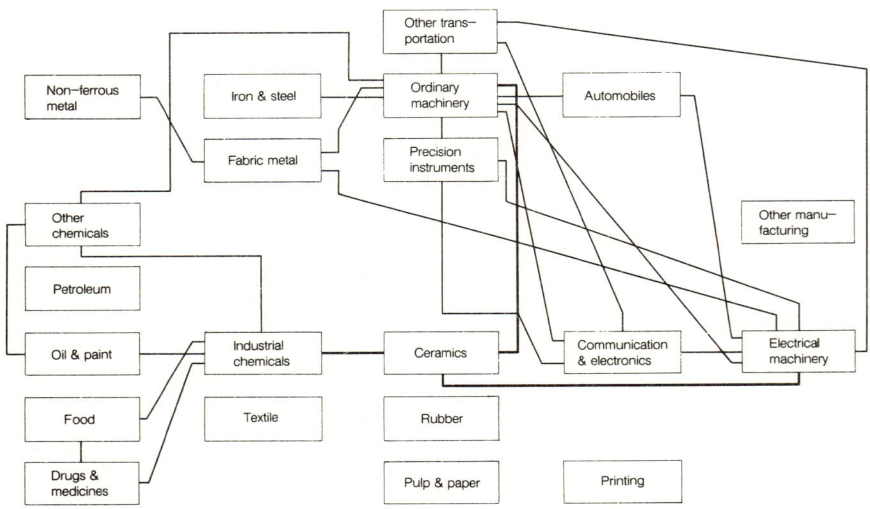

Figure 5.8 Technology fusion graph for new ceramics (in 1982)

We can make the following observations about the mega-trend in technology fusion from the 1960s to the 1980s – Before 1970, there was no major technology fusion. During the 1970s, fusions occurred only within the fabrication industries such as mechatronics and within the material industries such as biotechnology. However, there had been no fusion between the materials and the fabrication industries. Since 1980, fusions have appeared frequently between these two sets of industries which indicates the new innovation pattern of the so-called 'new materials revolution'.

Fortune [7] described the historical perspectives of this new materials revolution, as follows:

> The rise of civilization has depended on the emergence of materials that gave their names to whole epochs: the Stone Age, the Bronze Age, the Iron Age. Yet however vital they were to man's march from the cave to the computer, until recently the materials involved have been mostly nature's gifts or simple improvements on them: sand melted into glass, ores purified into metals. But now the world stands on the threshold of a new age of man-made materials.
>
> Today scientists can tailor the basic structures and properties of materials to suit their needs. Companies that lead in inventing and producing these ingredients of tomorrow will be in a strong position to dominate many high-technology industries.

By the same token, it is widely believed in Japan that the country is entering the age of fourth generation materials. First generation materials are stones and woods, which are used only by transforming natural resources. The second generation materials are copper and iron, which become usable by extracting components from the naturally available materials. A typical example of third generation materials are plastics, which are not available in nature but are made by synthesizing them artificially. Fourth generation materials will be designed according to usage by controlling the behavior of atoms and electrons in the materials, in the same manner as we design equipment and systems. This means that molecular structures can now be artificially manipulated to produce specific materials for specific uses.

Therefore, I would argue that the essence of the 'new material revolution' is the technology fusion between two industries, fabrication and materials, which had never been realized before. If this is the case, it will be no longer taken for granted that Japanese materials industries are much weaker compared to the fabrication industries. In fact, an early indication can be found in the results of the *Fortune* scoreboard in Table 5.1, which revealed that the ratio of Japan/US scores was as close as 0.82 in advanced materials, compared to 0.74 in computers and to 0.64 in life sciences.

An interesting implication derived from the findings that the new materials revolution is a product of technology fusion might be that the pattern of this revolution will not follow the conventional pattern of a materials revolution, i.e. the main actor of this revolution will not necessarily be the materials industry, but might be the fabrication industry, the customer for materials. We can find early indications of this possibility in o73 the recent Japanese development of optical fiber technology, in which NTT, the customer for optical fiber, developed the manufacturing technology, as described in Chapter 3.

This method of analysis has potential as a empirical technological forecasting method, which is totally different from conventional methods such as the Delphi Method, that rely heavily on the subjective human judgment of experts. If high technology is a product of fusion-type innovation, this can be the method of technological forecasting in high technology: reciprocal investment patterns as signs of future technology fusion can be used as leading indicators of high technology innovation. Because the forecasting can be done in an empirical way, this method has the potential to make forecasting more reliable.

7. Looking ahead

Discussions about high-technology industries often tend to center only around electronics. As far as the importance of materials is concerned, however, a growing consensus seems to be emerging among experts in various fields from various countries.

Tadahiro Sekimoto, president of the NEC Corporation, has pointed out, 'developments in the industry depend on developments in materials research'. Along the same lines, Professor Hiroshi Inose, winner of the Marconi Award for his work in digital communications, claims that the focus of technological development 'once shifted from systems to devices, and now it has shifted again to materials'. In the United States, Ralph E. Gomory, a senior vice-president and chief scientist at IBM, told a *Fortune* correspondent, 'every single step in computing has depended on solving one materials problem after another'. [7] As these comments indicate, expectations regarding advances in materials technology are greater now than ever before. Materials have clearly played an important role throughout history. How, then, are today's expectations of materials technology different from those of the past?

Roughly every five years, Japan's Ministry of International Trade and Industry draws up a list of major fields of technology. At the last such meeting, which I had the good fortune to chair, scientists and technicians representing seven technological fields gathered together to draft the latest list. This time, however, we tried a somewhat different approach.

We asked the participants from each field to list those fields, outside their own, from which they are most eagerly anticipating new advances. [22] We asked these questions of scientists and engineers who are working in corporate R&D and planning, in terms of their expectations in the short-term (0–5 years) and in the long-term (5–10 years).

The seven technical fields consisted of new materials, biotechnology, electronics, information processing, energy, aerospace and construction/transportation. Seventy scientists and engineers were sampled for each technical field: thus we sent the questionnaire to 490 people in 231 companies. As is shown in Table 5.5, a total of 149 responses were obtained, which was a return ratio of 30 percent. The most active response came from information processing, i.e. its return ratio was 42 percent, followed by energy (33 percent). This reflects the intrinsic nature of these technical fields, advances in which are heavily dependent upon advances in other technical fields.

Table 5.5 Number of respondents by field

Technological areas	Research/ development	Corporate planning	Total	Response ratio
new materials	15	6	21	30.0%
biotechnology	7	12	19	27.1
electronics	13	6	19	27.1
information	21	9	30	42.1
energy	18	5	23	32.9
aerospace	8	10	18	25.7
construction	11	8	19	27.1
total	93	56	149	30.4

Table 5.6 summarizes the results of our survey for the short term (0–5 years). The second to the bottom row gives each field's maximum possible number of responses, i.e. the total number of respondents minus the respondents in the field itself. Expectations are reported in two rows, the third to the bottom row recording the number of respondents in other fields who indicated high expectations for developments in that field, and the bottom row expressing this number as a percentage of the maximum possible number of respondents.

According to these scores and percentages, within the next 5 years, the greatest expectations are held for electronics, i.e. as many as 94 respondents in other fields, or 72 percent of all possible respondents, answered that they anticipate advances in that field. This is followed by new materials (85 points) and then by information processing (72 points). However, relatively little is anticipated from the fields of biotechnology, aerospace, energy and construction.

The high marks for electronics do not need any explanation. The high score for information processing reflects the fact that, like genetic engineering, it is expected to provide important tools for research and analysis. Expectations *vis-à-vis* biotechnology remain low, however, as this area has yet to move much beyond the conceptual stage.

The low expectations regarding aerospace and energy, however, merit particular attention. It is widely known that state-sponsored aerospace and energy projects, many of which are military-related, have produced numerous breakthroughs in new materials, electronics and software, in the form of spinoffs. The low expectations for this field, however, seem to indicate that the development of new materials and computers is generally no longer expected to follow these conventional mechanisms. New materials and computers are now expected to be developed in direct

Table 5.6 Expectations for development across technological fields

Expectations by	Expectations of						
	New materials	Biotechnology	Electronics	Information	Energy	Aerospace	Construction /trans
New materials	–	2	15	10	11	9	2
Biotechnology	12	–	12	14	0	1	1
Electronics	16	1	–	13	4	6	2
Inf. processing	11	3	29	–	2	2	1
Energy	19	2	18	14	–	3	6
Aerospace	13	3	11	9	4	–	2
Const./trans.	14	3	9	12	6	2	–
total (A)	85	14	94	72	27	23	14
no. of possible respondents (B)	128	130	130	119	126	131	130
(A/B)	66%	11%	72%	61%	21%	16%	11%

response to specific technological needs, rather than emerging in a roundabout way from defense-related projects.

Table 5.7 summarizes the results for the short term (0–5 years) and the long term (5–10 years), so that the shift from the short to the long term can be seen. In the longer term, great expectations are held for new materials, i.e. 94 positive responses or 73 percent of all possible responses. Expectations for electronics fall to the second place in the longer term, 78 points or 60 percent. This reflects the widely-held view that the ultimate problem-solving will be effected only by the development of new materials, not by application of electronics.

Increasing expectations regarding new, directly applicable materials technology reflects, at least in part, a dramatic change in the nature of technological innovation in the materials field. The terms 'materials design' and 'custom materials' have also recently entered the scientific lexicon. As described before, this means that molecular structures can now be artificially manipulated to produce specific material for specific uses.

It is hoped that, in the future, biological processes will come to replace physical and chemical processes in various fields of manufacturing. This is reflected in the increased expectations for biotechnology in the long-term view (5–10 years). Its score increases from 14 points (11 percent) in the short term to 62 points (48 percent) in the long term.

Table 5.7 Expectation of advances in each technological area

	Possible no. of responses	Short term (0–5 years)		Long term (5–10 years)	
	A	B	B/A	C	C/A
new materials	128	85	66%	94	73%
biotechnology	130	14	11%	62	48%
electronics	130	94	72%	78	60%
information	119	72	61%	68	57%
energy	126	27	21%	45	36%
aerospace	131	23	16%	38	29%
construction	130	14	11%	16	12%

8. Synthesizing techno-paradigm shifts

By conntrasting high technology with modern technology, we ascertained that technology fusion is an important characteristic of high technology innovation. Technology fusion was also contrasted with technical breakthroughs, and a number of revealing comparisons were made between technical innovations achieved through technology fusion and those achieved through technical breakthroughs.

We found that the essence of technology fusion is reciprocal R&D

investment. It was also argued that reciprocal research investment is more than simply an analytical tool for studying technology fusion. Because technology fusion is more than a combination of different technologies, it is unlikely to be attained through merger and acquisition; it definitely needs joint research among different companies in different industrial sectors.

The quantitative analysis clearly suggested that the technical innovations involved in high technology are better described as technology fusion than as technical breakthroughs. Expressed more accurately, however, a single technical breakthrough alone is not sufficient for progress in high technology. Rather, only through the organic fusing of several technical breakthroughs in a number of different fields can a new technology be created.

In what follows, I will try to synthesize those categories of techno-paradigm shift described in preceding chapters around the concept of technology fusion. In other words, the shift of innovation patterns toward technology fusion is a trend underlying the other categories of techno-paradigm shift.

The relationship between technology fusion and manufacturing companies' becoming thinking organizations is observable. Technical terms are increasingly being used as catch-phrases for corporate identity and for defining a corporate business domain. For example, C&C (computer and communication) is used by NEC, and E&E (energy and electronics) by Toshiba. As those phrases imply, technology fusion is envisioned clearly, and it is reported that such phrasing helped in shifting these companies into a growing area. As described earlier, the recent rapid increase in R&D investment, which is surpassing capital investment, partly originates in a widening of research programs as they encompass more areas. The shift of innovation pattern toward technology fusion might be one driving force behind the shift toward thinking organizations and will probably accelerate it further in the future.

Technological diversification is at least a necessary condition for technology fusion. It leads to technology fusion, because technological diversification in Japan is attained through diversification of R&D. Through the technological diversification effort already made, Japanese companies have built the fundamental basis for technology fusion. Further diversification might lead to the situation in which it becomes possible for a single organization to realize technology fusion.

The techo-paradigm shift in R&D activities will facilitate the realization of technology fusion. One corporate strategy for insuring themselves against the possibility that their technical problems will be solved by companies outside their industrial sector, is to form technical alliances with these companies. These alliances between companies in different industrial sectors work not only as a competitive hedge against technological surprises which might be brought about by companies in different industrial sectors, but also as a device which facilitates technology fusion.

Technology fusion is intrinsic to the process of demand articulation, because demand articulation is defined as the search and selection process among technical options. When component technologies are not available within existing technical collections, a long-term technology development effort is needed. However, when demand is well articulated, development activity can be made complimentary to the other technologies being developed. In other words, technological development activities through demand articulation drive technology fusion, or sometimes force it.

The shift to technology fusion is well reflected in the recent changes in the way in which collective research is organized. The analysis of membership in research associations reveals that the average number of industrial sectors per project in the research consortia is increasing, while the number of participating companies per industrial sector is decreasing. In other words, collective research is beginning to combine firms in different industrial sectors rather than different companies within the same industrial sector. In fact, based on data about the industrial sectors of participating companies, quantitative analysis has been made of technology fusion. [23] Similar results have been produced in terms of mechatronics, macro-system technologies, new materials and biotechnology. A similar result is also reported in an analysis made of joint patent applications among different companies. [24]

So we see that the changing focus of manufacturing companies, the diversification of R&D, the changing pattern of R&D activities, the increasing importance of demand articulation and the nature of cooperative research activities are all related to the increasing importance of technology fusion in creating technologies.

In the next chapter, I will move from examining production of technologies to analyzing diffusion of technologies.

References

[1] *Newsweek*, 'Can the Japanese Create?', 2 July 1984, p. 27.

[2] S. Feiman and W. Fuentevilla, *Indicators of International Trends in Technological Innovation*, Gellman Associates, April 1976.

[3] F. Narin and J. Frame, 'The growth of Japanese science and technology', *Science*, vol. 245, pp. 600–605, 1989.

[4] N. Rosenberg, *Inside the Black Box: Technology and economics*, Cambridge University Press, 1983.

[5] MITI, *Vision for Industrial Machinery* (in Japanese), p. 220 Tsushou-Sangyou-Chousakai, Tokyo, 1984.

[6] B. Johnstone, 'Mechanics meets electronics', *New Scientist*, vol. 110, no. 1503, pp.35–38, 1986.

[7] G. Bylinsky, 'The high tech race: who's ahead?', *Fortune*, 13 October, pp. 18–36, 1986.

[8] N. Rosenberg, *Perspective on Technology*, p. 28, Cambridge University Press, 1976.

[9] Japan Industrial Policy Research Institute, *A study of the Innovation Process* (in Japanese), Sangyou-Kenkyusho, Tokyo, July 1989.

[10] N. Rosenberg, 'Technological change in the machine tool industry, 1840–1910', *Journal of Economic History* vol. 23, no. 4, 1963.
[11] F. Kodama, 'Japanese innovation in mechatronics technology: a study of technological fusion', *Science and Public Policy*, vol. 13, no. 1, pp. 44–51, 1986.
[12] C. Freeman, *Technology Policy and Economic Performance*, pp. 87–8, Pinter Publishers, 1987.
[13] F. Kodama, 'Can empirical quantitative study identify changes in techno-economic paradigm?', in *Science, Technology and Industry Review*, no. 7, pp. 101–129, OECD, Paris, July 1990.
[14] J. Tilton, *International Diffusion of Technology*, The Brookings Institution, 1971.
[15] R. Levin, 'The semiconductor industry', in *Government and Technical Progress: A cross-industry analysis*, R. Nelson (ed.), Pergamon Press, 1982, pp. 9–100.
[16] OECD, *Case Study of Electronics with Particular Reference to the Semiconductor Industry*, Joint Working Party of the Committee for Scientific and Technological Policy and the Industry Committee on Technology and the Structural Adaptation of Industry, pp. 133–63, November 1977.
[17] N. Rosenberg, 'Civilian 'spillovers' from military R&D spending: the American experience since World War II', in *Technical Cooperation and International Competitiveness*, H. Fusfeld and R. Nelson, (eds.), Rensselaer Polytechnic Institute, 1988, pp. 167–87.
[18] MITI, *Electronics & Machinery Industry in the Seventies* (in Japanese), p. 50, Tsushou-Sangyou-Chousakai, Tokyo, 1971.
[19] Statistical Bureau, Prime Minister's Office of Japan, *Report on the Survey of Research and Development*, Nihon-Toukei-Kyoukai, Tokyo, 1970–82.
[20] F. Kodama, 'Technological diversification of Japanese industry', *Science*, vol. 233, pp. 291–6, 1986.
[21] M. Dibner, 'Biotechnology in pharmaceuticals: the Japanese challenge', *Science*, vol. 229, p. 1230, 1985.
[22] Economic Research Institute, *A survey on the interaction among Important Industrial Technologies* (in Japanese), Japan Society for the Promotion of Machinery Industry, Tokyo, May 1990.
[23] I. Shirai and F. Kodama, *Quantitative Analysis on Structure of Collective R&D Programs by Private Corporations in Japan* (in Japanese), NISTEP Report no. 5 (National Institute of Science and Technology Policy, Tokyo, 1989).
[24] K. Kotani, 'Measurement of innovation process by output indicators: a measurement of technology fusion and of innovation spiral by patent information', (in Japanese), MS dissertation submitted to Graduate School of Policy Science, Saitama University, 1988.

Chapter 6
Technology diffusion: from technical to institutional inertia

1. Introduction

In his taxonomy of technical change, Freeman [1] specifies four categories: incremental innovations; radical innovations; changes of technology system; and changes in techno-economic paradigm. He states:

Some changes in technology systems are so far-reaching in their effects that they have a major influence on the behavior of the entire economy. The diffusion of steam power and of electric power are obvious examples of such deep-going transformations. So too, is the combination of innovations associated with the electronic computer. Such changes in paradigm make possible a 'quantum leap' in potential productivity which, however, is at first realized only in a few leading sectors. In other sectors such gains cannot usually be realized without organizational and social changes of a far-reaching character.

On the basis of the above description, we can formulate the problem as follows: the change in social institutions is in fact the determinant of the diffusion of the techno-economic system. More specifically, the innovation which involves substantial change in organization if it is to be exploited diffuses more slowly than the innovation which does not need such changes.

However, there exists a surprisingly small amount of empirical and quantitative evidence regarding such a techno-economic paradigm change, although such changes are mentioned by several authors in general ways.[2] Therefore, the purpose of this chapter is to conduct empirical and quantitative studies on the diffusion process of techno-economic paradigm changes.[3]

This chapter consists of two parts: descriptive evidence and quantitative evidence. The descriptive evidence is drawn from two case studies in the fields of telecommunication and welfare. We chose these two fields because they are subject to heavy regulation, and so we are able to examine how changes in regulations will interact with the diffusion of technology.

In the second part of this chapter a quantitative analysis is made of the diffusion of computer systems into local government. This sector was chosen for two reasons. First, because it is obviously not a leading sector in computer utilization, the diffusion in this sector is most appropriate for the definition given by Freeman. Secondly, a unique database is available in Japan, which allows quantitative study on the organizational aspect of diffusion of a new techno-economic system.

2. Descriptive evidence: case studies of facsimiles and CAT scanners

In the field of telecommunication, institutional changes can often trigger the diffusion of technology. In other words, the changes in regulation result in the initiation of the diffusion process. One example is the progress being made in developing new technologies for facsimile communications in Japan.[4]

Given the particular character of the Japanese language, potential demand for this type of communication has surfaced, taking advantage of the liberalization of the use of telephone circuit networks through the revision of the telecommunication law.

The Revision of the Public Telecommunications Law and the consequent opening of the telephone network for free use in 1971 permitted the emergence of latent demands for facsimile communications: the facsimile itself being inherently suited to the Japanese language. The market that had hitherto been limited to special applications and specific businesses was now wide open to general business, and research and development in both terminal and network technologies was launched.

In the field of welfare-oriented products, the technology has to be diffused within the framework of the existing national welfare system. However, the introduction of new technology sometimes induces the changes in regulation, which is the case in the development of CAT scanners.[4]

The first CAT scanner was developed by EMI of the United Kingdom, but its diffusion rate was low in Japan. When the number of CAT scanners surpassed 200 units, a professional association of experts was established. This association demanded that the Ministry of Health should permit the use of scanners to be covered by the national health insurance scheme. The government decided to do so, and afterwards the market of scanners expanded rapidly.

2.1 Diffusion of facsimile communication

As of March 1988, the worldwide installation of facsimiles has reached 4.3 billion units, and 52 percent of them are installed in Japan, as shown in Table 6.1.

A rough estimate of the diffusion level as of March 1989 is that the number of facsimiles installed in Japan amounts to 6 percent of the number of the total telephone lines and to 24 percent of business lines. In other words, it is equivalent to the situation in which a facsimile is

Table 6.1 Number of installed units of facsimiles by regions (thousand units)

Year	Total	Japan	N. America	Europe	Others
1987	3,250	1,800 (55%)	800 (25%)	400 (12%)	250 (8%)
1988	4,250	2,200 (52%)	1,100 (26%)	600 (14%)	350 (8%)

Notes: 1: The world share is in parenthesis; 2: The 1987 data is units as of September 1987, and 1988 data is as of March 1988.

Source: Nikkei Communications, 28 March, p.46, 1988.

available to every four business persons if we assume that one telephone line is used by one business person. This level is almost a saturation, as far as the use of facsimiles for business purposes is concerned. Therefore, future expansion might lie in home use.

It is also surprising how rapidly this saturation level has been reached. In Table 6.2, the time series of annual production of facsimiles is shown. As can be seen in the Table, annual production has grown by more than forty times in less than ten years, which indicates that potential demand for this type of communication could have been tremendous in Japan.

Table 6.2 Time-series of annual production of facsimiles

Year	Number of units (thousand)	Value in yen (¥ billion)
1980	100	81.1
1981	127	105.6
1982	229	143.5
1983	304	168.8
1984	511	247.8
1985	886	313.2
1986	1,234	303.0
1987	2,414	364.8
1988	4,328	465.9

Source: MITI.

According to a sample survey conducted by the Japan Management Association, the facsimile has been more widely adopted than other popular office equipment such as word processors, high speed PPCs (plain paper copiers) and personal computers, as shown in Table 6.3. The table also reveals that there is no difference between small offices and large offices in the adoption rate of facsimiles, while there is a fairly big difference for other office equipment. This situation is reflected in the historical development of facsimile communication in Japan.

As early as 1930, the Ministry of Communications, now the Ministry of Posts and Telecommunications, started the telephotography service,

Table 6.3 Adoption rate of office equipment (1988)

	Small offices	Large offices
facsimile	96.0%	96.8%
word Processor	81.3	93.5
PPC(high speed)	85.7	90.3
personal computer	71.8	95.3

Note: Small office is defined either as the office of companies in which the number of employees is between 50 and 300, or as the city office of the population of less than 100,000.
Source: *White Paper of Office Automation*, Japan Management Association (Nihon-Keiei-Kyoukai), April 1990.

which was a forerunner of today's facsimile machines, between Tokyo and Osaka. During that period, however, the primary use of the service was for special applications, such as transmitting the photographs and diagrams of the news media and weather service. Subsequently, the police, as well as large banks, and electric and railroad companies, which needed to transmit massive amounts of documents, began to use the service.

In the years up to 1965, use by general business corporations expanded, and local municipalities also began to use the service for various tasks such as census registrations. However, the spread of the service was limited because terminal equipment was structurally complex and expensive as were communication fees. Moreover, usable lines were then limited to non-switched lines only, which were leased to subscribers and thus charged at a fixed rate. The situation was identical in other countries, so the facsimile was often referred to as a 'sleeping giant'.

Partial amendment of the Public Telecommunications Law in May 1971 allowed the telephone network to be used for purposes other than voice transmission. Although facsimile use was previously limited to non-switched leased lines, deregulation of the network enabled facsimile equipment to be connected to subscriber lines. This meant that charges could be levied according to the length of time and distance used for communication, and not at a fixed rate. Furthermore, any facsimile could freely communicate with another facsimile on the public telephone network if they could be properly interfaced. Thus with the freeing of the network facsimile communications began to spread rapidly.

In 1973, a new public telecommunications service was inaugurated. This was a facsimile service, and was connected to the switched telephone network and enabled a subscriber to connect a facsimile to his line. Communication equipment manufacturers simultaneously began to sell equipment to telephone subscribers. One company after another began to use the service for general communications, that is, to send documents between offices, and usage spread from large to small businesses and stores. Less than two years after deregulation, the number of facsimiles installed doubled, from 8,800 units in 1971 to 16,000 units in 1973.

The rapid growth in facsimile use was accompanied by rapid progress in scanning and recording techniques. This was due to intensified competition among equipment manufacturers. At the same time, transmission time was reduced through the development of band-width compression techniques. Furthermore, efforts were made to develop low-cost, high reliability, compact facsimiles for household use. Moreover, studies were begun on a 'facsimile communications system' to be integrated with a stored and a forward-switching network providing versatile services at a low communications costs. This system was inaugurated in September 1981.

The Japanese language induced the early development of the facsimile, and its diffusion in the rest of the world came later, as shown in Table 6.4. Because of various technology developments made in Japan, facsimile communication surpassed its predecessor, the telex, in every functional aspect. Therefore, it also diffused very rapidly in countries which have different traditions and communicate in different languages. It should also be noted that the diffusion paths in United States and Europe are following exactly the same path in spite of the differences in business and culture between these two regions. This fact indicates how benefits brought about by facsimiles are universal throughout the advanced countries.

Table 6.4 Installation of facsimiles in the United States and Europe (thousand units)

Year	USA	Europe
1983	75	-
1984	124	133
1985	251	199
1986	400	360
1987	696	739
1988	1,641	1,662
1989	2,849	2,991

Source: Dataquest.

However, a careful look at adoption rates by size including smaller offices of less than fifty employees, reveals that the size is a critical determinant of the diffusion even in case of this universal machine. Adoption rates of facsimiles by office size are shown in Table 6.5. As seen in the table, the adoption rate by offices of less than 10 employees surpassed 50 percent only in 1988, although it was almost 100 percent in offices of more than 300 employees. Therefore, even in the technology which is widely accepted all through different cultures, we can conclude that size is a critical factor in technology diffusion.

Table 6.5 Adoption rate of facsimiles by size (as of 1988)

Number of employees	Percent of adoption
–300	99.0
100–299	92.7
50– 99	90.8
20– 49	83.3
10– 19	73.1
5– 9	55.2

Source: Data Communication, p. 43 (in Japanese, Dehta-Tsusin), May 1990.

2.2 Diffusion of CAT

The CAT scanner is a highly sophisticated imaging modality that produces cross-sectional pictures of internal organs of the body, using a special configuration of X rays, detectors and computers.

The first X-ray CAT scanner was developed by EMI of the United Kingdom, with the cooperation with the British Department of Health and Social Security. CAT scanners were expensive, costing as much as 200 million yen, and therefore diffused slowly in Japan. The Japanese government purchased only one unit for the Research Institute of Legal Medicine.

However, since 1976, the Japanese government was trying to solve the problem of trade surplus. When it was decided that Queen Elizabeth would visit Japan, the Japanese government decided to purchase thirty-three CAT scanners, at a price of ¥200 million each.

The idea came from the Automotive Insurance Association. At that time, ownership of passenger cars increased dramatically in Japan; many traffic accidents caused many cases of brain damage which were expensive for the insurance agencies. In other words, it was very important for insurance companies to ascertain whether there was brain damage or not.

These purchased CAT scanners were distributed to brain surgery departments in major public hospitals: it was the first time many doctors had a chance to use the scanners. In medical schools, doctors learn medicine through cross-sectional pictures of the human body, but in practice they could only view pictures of vertical section, which were provided by X-ray equipment. After being able to use the scanners, therefore, many doctors felt they could not do without them. As a result, Japan became a country with many scanners.

When the number of CAT scanners surpassed 200 units, the number of experts reached 1,000, because 5 experts are needed on average per CAT scanner. A professional association of experts was established. This association became a lobbying group demanding that the Ministry of

Health permit the use of scanners to be covered by the national insurance scheme. In November 1978, the government decided to cover the use of scanners in the insurance scheme, and the market for scanners expanded rapidly.

Although Japanese medical equipment manufacturers underestimated the market potential for CAT scanners in the early stages, they soon recognized the profit potential, initiated research and development efforts and eventually built a mass production system. Thus the average price of a CAT scanner was reduced to one-fourth of its original price, i.e. 50 million yen. In 1988, Japanese production of CAT scanners reached ¥76.3 billion, and exports reached ¥33.3 billion.

As of 1988, approximately 6,000 units had been installed in Japanese hospitals. There are two types of CATs: one is a head scanner for the brain only and the other is a body scanner for the whole body, as shown in Table 6.6. Although the early scanners were for the head only as many as three-quarters of CAT scanners installed are body scanners, as seen in the table. And 90 percent of them in both types are made by Japanese manufacturers.

Table 6.6 Number of CAT scanners installed in Japan (as of August 1988)

	Head	body	total
Japanese made (A)	1,403	4,072	5,475
foreign made	110	511	621
total (B)	1,513	4,583	6,096
share (A/B)	93%	89%	90%

Source: *New Medicine* (Shin-Iryou), p.75, October 1988.

Diffusion rates of CAT body scanners in hospitals are shown in Table 6.7. In the table, the percentage of adoption is reported by hospital size. In 1989, the adoption rate reached 100 percent in hospitals with more than 500 beds, while it remained less than a quarter in those with less than 100 beds. And the estimate of the adoption rate in all hospitals reached 53.6 percent in 1989. The pattern of adoption reveals that adoption rates have reached almost saturation point in hospitals with more than 200 beds, while the rate for hospitals with less than 200 beds is still growing. Therefore, the time path of adoption rate in all the hospitals is still following that of exponential growth.

In his book on the economic analysis of CAT scanners in the United States, M. Trajtenberg [5] said that, although CAT scanners originated in England, the United States almost immediately became the prime locus of innovative activity in this area, primarily reflecting the fact that it accounts for a full half of the world market for imaging technologies.

Table 6–7 Diffusion of body CAT scanners by hospital size

No. of beds	No.of hospitals (total)	(sample)	% hospitals with CAT body scanners of 1982	1983	1984	1985	1986	1987	1988	1989
– 99	4,551	177	–	1.1	2.1	2.5	1.7	7.6	11.9	23.5
100–199	2,203	212	6.4	17.1	19.8	27.4	40.3	50.2	59.4	72.2
200–299	980	180	24.0	31.4	43.1	51.9	68.1	79.7	90.6	95.6
300–399	539	149	45.5	57.4	67.8	78.8	84.9	92.1	96.0	97.9
400–499	255	95	58.1	68.8	79.5	90.9	96.4	97.9	97.9	98.9
500–	412	102	72.5	81.8	88.6	92.1	97.9	98.1	99.0	100.0
All	8,940	915	10.3	15.4	19.2	23.5	29.0	37.0	43.8	53.6

Note: A hospital is defined as the medical institution which has the capacity for more than twenty patients. However, hospitals for psychiatry, tuberculosis and leprosy are not included in this table.

Source: Ministry of Health, for number of hospitals. For the diffusion of CAT, Japan Hospital Association, 'Reports on Hospitals', (Byouin-Gaikyou-Chyousa-Houkoku). For the diffusion data in 1988, The Japan Hospitals Federation, 'Management analysis survey of hospitals in Japan' (Byouin-Keieibunseki-Chyousahoukoku).

In order to show the diffusion rate of CAT scanners in Japan, figures for Japan are compared with corresponding US data reported in Trajtenberg's book. However, the US data is only up to 1984 and do not include hospitals with less than 100 beds. Moreover, his data do not make a distinction between head and body scanners. In the Japanese data, the combined adoption rates of head and of body scanners is calculated, although there is a danger of duplication. The comparison is shown in Table 6.8.

Table 6.8 US – Japan comparison in diffusion rate of CAT scanners (1984)

Number of beds	% of hospitals with CAT		
	Japan (A)	USA (B)	(A/B)
100–199	28.8%	29.2%	0.98
200–299	62.2	64.0	0.97
300–399	86.1	84.6	1.02
400–499	96.2	97.1	0.99
500–	100.0	100.0	1.00
All	53.9	57.0	0.94

As can be seen in the table, in the all the hospitals with more than 100 beds, the Japanese adoption level is equivalent to 94 percent of the US level. However, when it comes to a size-by-size comparison, there is almost no gap between the two countries. Therefore, the difference in total comes mainly from the difference in size distribution of hospitals between the two countries. It should be noted that the Japanese rate surpasses the US rate in middle-sized hospitals of 300–399 beds.

Finally, in order to show the size effect, the differences in adoption rate of body scanners among different sizes of hospitals are shown in Table 6.7. As can be seen in the table, size is a critical determinant during the diffusion period, i.e. until 1986, although it is less critical at the end of the period.

3. Database for analysis of quantitative evidence

In this section, we will describe a set of fairly unique Japanese databases, which facilitate quantitative studies of the techno-economic paradigm shift. Let us specify the requirements for the database to accommodate the organizational aspect of diffusion of a new techno-economic system. Metcalfe [6] makes the distinction between adoption and diffusion. He writes:

Adoption analysis considers the decision taken by agents to incorporate a new technology into their activities. It leads to propositions linking the nature and timing of adoption decisions to specified characteristics of adopters.

By contrast, *diffusion analysis* is concerned with how the economic significance of a new technology, which is often measured by the share of the market held by a product innovation or by the fraction of industry output produced with a process innovation, changes over time. In this sense the analysis of diffusion is closely related to the analysis of technological substitution in which a 'new' technology displaces an 'old' technology.

The diffusion of a techno-economic system is more than just technological substitution. Therefore, adoption analysis rather than diffusion analysis is necessary. However, adoption analysis requires the following database:

1. the existence of a substantial number of adopting agents;
2. the identity in mission among the adopting agents;
3. independence in decision-making among agents; and
4. the heterogeneity of adopter characteristics.

Japan is divided into forty-seven administrative units, called prefectures. Prefectural governments are homogeneous by nature and similar in organizational principle and missions. Their tasks are specified in detail by the law of prefectural government.

However, they are heterogeneous in terms of size and decision-making. Since the division of Japan into forty-seven prefectures is based on land area and not on population size, they are very different in terms of size as measured by population, as well as by budgets. As far as decision-making is concerned, in the case of ministries and agencies of the central government their budget is controlled by the Ministry of Finance, through budget allocation. On the other hand, since there is no such overall control over prefectural governments, they can be assumed to be independent in terms of decision-making.

Moreover, it is an absolute necessity to have the information concerning how and for what purpose the technology is used, in order to accommodate the organizational aspects of technology diffusion. This is very important in the case of the computer, because some utilization might involve a substantial change in organization, and some might not.

The database collected by the Ministry of Home Affairs could accommodate these requirements. [7] As far as the date of computer installation is concerned, information has been compiled every year since 1969, as shown in Table 6.9.

As to the utilization of the computer, the application is divided into forty-one task categories, such as payroll calculation, taxation, planning, and so forth. In other words, the survey asks whether the installed computer is being used for each of the forty-one task categories or not. This database about computer utilization has been available every year since 1963. Time-series data of diffusion rates of some typical task categories are shown in Table 6.10.

Table 6.9 Diffusion of computer installation in prefectural governments

Year	No. of prefectures	Diffusion rate
1963	1	0.02
1964	2	0.04
1965	4	0.09
1966	5	0.11
1967	7	0.15
1968	12	0.26
1969	14	0.30
1970	20	0.43
1971	23	0.49
1972	27	0.57
1973	39	0.83
1974	43	0.91
1975	44	0.94
1976	45	0.96
1977	46	0.98
1978	47	1.00
1979–	47	1.00

Note: The definition of computer is in accordance with that established by MITI and JECC, stating it should be digital processing, be programmable, have more than 2,000 bits memory and electronic logical calculation. [8]

4. Choice of mathematical models

In this section, the mathematical models used in this chapter are described. These were chosen from the existing diffusion models because they are designed to accentuate particular aspects of a new techno-economic system.

A number of authors have made important contributions to the literature on diffusion. According to Stoneman,[9][10] there are two kinds of theories about intra-sectoral diffusion: the psychological approach; and the probit approach. Stoneman writes.

> The Mansfield model [11] is 'psychological' in the sense that the speed of response is assumed to be related to the size of the stimulus. It is also epidemic in the sense of learning by infection. In the probit approach the reason why all firms do not use the new technology instantaneously is that it is not profitable for all firms to do so.

In the psychological approach, the assumption about the agent's adopting decision is that uncertainty is reduced in proportion to the number of agents already using the new technology. Therefore, a logistic curve is obtained to suggest the diffusion pattern. Thus, we can say, the implicit assumption behind the model is that information spreading mechanisms are based on personal contact. [12]-[16]

Table 6.10 Diffusion rates of computer utilization for task categories

Year	Personnel management	Payroll calculation	Taxation for auto mobiles	Taxation for corporations	Management of issuing bonds	Production of designated statistics	Fore-casting & planning	Forest planning	Cost estimation of civil works
1963	0.04	0.07	0.02	0	0.02	0.04	0	0	0
1964	0.04	0.07	0.02	0	0.02	0.07	0	0	0
1965	0.04	0.15	0.04	0.02	0.02	0.11	0	0	0
1966	0.04	0.17	0.04	0.02	0.02	0.13	0	0	0
1967	0.04	0.20	0.07	0.02	0.02	0.13	0	0	0
1968	0.09	0.24	0.15	0.07	0.02	0.17	0.02	0.04	0
1969	0.13	0.28	0.24	0.07	0.02	0.26	0.09	0.09	0
1970	0.26	0.41	0.33	0.07	0.02	0.30	0.13	0.11	0
1971	0.33	0.48	0.39	0.11	0.09	0.37	0.13	0.13	0.04
1972	0.46	0.54	0.46	0.15	0.20	0.41	0.24	0.20	0.04
1973	–	0.55	0.55	0.19	0.21	0.43	0.30	0.36	0.04
1974	–	–	0.62	0.28	–	–	0.43	0.53	–
1975	–	–	0.62	0.28	–	–	0.51	0.57	–
1976	0.81	0.94	–	–	–	–	0.66	0.94	0.51
1977	0.77	0.81	0.75	0.38	0.57	0.75	0.55	0.60	0.45
1978	0.79	0.79	0.72	0.40	0.66	0.75	0.60	0.64	0.51
1979	0.79	0.79	0.75	0.45	0.64	0.81	0.64	0.62	0.60
1980	0.79	0.81	0.75	0.51	0.72	0.83	0.68	0.64	0.62
1981	0.79	0.81	0.77	0.60	0.77	0.83	0.70	0.64	0.64
1982	0.85	0.85	0.79	0.66	0.79	0.87	0.70	0.72	0.77
1983	0.85	0.85	0.81	0.75	0.79	0.87	0.75	0.75	0.85
1984	0.85	0.87	0.79	0.75	0.81	0.85	0.70	0.75	0.77
1985	0.85	0.87	0.81	0.79	0.83	0.85	0.70	0.75	0.83

Source: Ministry for Home Affairs. [7] *Note*: – = not available.

In the probit approach, on the other hand, the agent's adopting decision is assumed to be based on profitability. However, in the government sector, it is difficult to estimate profitability in a direct way. Therefore, the size might have a more direct relevancy to the adopting decision. In fact, through the two case studies made in the preceding sections we found that size is a critical determinant of the diffusion, even in the areas outside governmental sectors. Therefore, the mathematical model which deals with size in an explicit way, will be chosen among various existing models which belong to probit approach.

4.1 Psychological approach

The basic hypothesis of the simple epidemic model is represented by the following equation:

$$m_{t+1} - m_t = \beta \cdot (n - m_t) \cdot m_t / n, \ \beta > 0,$$

where m_t is the number of individuals, in a fixed population of n, having contracted an infectious disease at time t. Thus the number of individuals contracting the disease between times t and $t+1$ is proportionate to the product of the number of uninfected individuals and the proportion of the population already infected, both at time t.

If the period t to $t+1$ is very small, it may be alternatively stated as:

$$d/dt[m_t/(n-m_t)] = \beta \cdot (m_t/n).$$

This differential equation has the following solution:

$$m_t/n = \{1 + \exp(-\alpha - \beta \cdot t)\}^{-1}, \quad (1)$$

where α is a constant of integration.

Therefore, when a diffusion path follows the logistic curve, we can assume that the determinant of diffusion is the speed of information-spreading based on personal contact.

4.2 Probit approach

The probit approach, on the other hand, is based on the more explicit behavioral assumption. David [17][18] writes:

> Whenever or wherever some stimulus variate takes on a value exceeding a critical level, the subject of the stimulation responds by instantly determining to adopt the innovation in question. The reason why such decisions are not arrived at simultaneously by the entire population of potential adopters lies in the fact that at any given point of time either the 'stimulus variate' or 'critical level' required to elicit an adoption is described by a distribution of values, not a unique value appropriate to all members of the population. Through some exogenous or endogenous process, however, the relative position of stimulus variate and critical response level are altered as time passes, bringing a growing proportion of

the population across the 'threshold' into the group of actual users of the innovation.

As described before, search efforts have been made to discover the mathematical model in which size is dealt with explicitly. Davies [19] builds his diffusion model around the assumption that an agent will use a new technology if the expected pay-off period from its use becomes smaller than some critical pay-off period. The expected pay-off period is a function of agent size and time. And the critical pay-off period is also related to agent size. In other words, both the expected pay-off period and the critical period vary with agent size. On the basis that actual agent sizes are lognormally distributed, Davies developed a theory of diffusion.

Let

$ER_i(t)$ = agent i's expectation of the pay-off period associated with adoption at time t, and
$R_i^*(t)$ = agent i's critical pay-off period against which adoption is assessed at time t,

then, an agent i will adopt the innovation by the time t if

$$ER_i(t) \leqslant R_i^*(t).$$

The expected pay-off period, ER, is then a function of agent size S_i, time factor $\theta_1(t)$, and other characteristics ϵ_i^1:

$$ER_i(t) = \theta_1(t) \cdot S_i^{\delta} \cdot \epsilon_i^1,$$

where

$$\epsilon_i^1 > 0,$$
$$\theta_1(t) > 0.$$

We can assume that ER will decline monotonously with time for all i, because of successively improved later vintages of the process resulting from learning by doing by the innovation suppliers and of the strong likelihood that incoming information resulting from adopting agents' search will tend to improve their views of the profitability of adoption. Therefore,

$$\{d\theta_1(t)/dt\}/\theta_1 < 0 \text{ for all } t. \tag{2}$$

The critical pay-off period, R^*, is also related to firm size, time factor and other characteristics:

$$R_i^*(t) = \theta_2(t) \cdot S_i^{\mathrm{r}} \cdot \epsilon_i^2,$$

where

$$\epsilon_i^2 > 0,$$
$$\theta_2(t) > 0.$$

By the same token, R^* should increase monotonously with time for all i. It reflects relaxations in the risk premium and the effect of competitive pressures. Therefore,

$$\{d\theta_2(t)/dt\}/\theta_2 > 0 \text{ for all } t. \tag{3}$$

The ownership condition can be rewritten: $ER_i(t)/R_i^*(t) \leqslant 1$. Since both the expected pay-off period and the critical period vary with agent size, the ownership condition can be again rewritten as:

$$ER_i(t)/R_i^*(t) = [\theta(t)\cdot S_i^{\beta}\cdot\epsilon_i]^{-1} \leqslant 1,$$

where,

$$\beta = \tau - \delta,$$
$$\epsilon_i = \epsilon_i^2/\epsilon_i^1 > 0,$$
$$\theta(t) = \theta_2(t)/\theta_1(t) > 0.$$

And, it follows from (2) and (3) that:

$$\{d\theta(t)/dt\}/\theta > 0 \text{ for all } t,$$

We can also assume that $\beta > 0$, because larger agents will have a greater probability of ownership at any point of time. Thus the ownership condition may be again re-expressed as:

$$S_i \geqslant S_i^*(t),$$

where $S_i^*(t)$ is referred to as critical agent size, and

$$S_i^*(t) = [\theta(t)\cdot\epsilon_i]^{-1/\beta}.$$

The multiplicative form of the central limit theorem suggests that ϵ_i will be lognormally distributed across agents: log ϵ is normally distributed across agents with mean zero and variance σ^2. Therefore, given a distribution of agent size, then, we have a theory of diffusion.

Let
$Q(t)$ = the probability that an agent chosen at random will have adopted by time t,
then
$Q(t) = Probability\ [S \geqslant S^*(t)]$.

5. Two types of innovations

5.1 Davies's distinction

As far as the pay-off period of new technology is concerned, Davies makes the following distinction: Group A innovations are relatively cheap and simple; Group B innovations are expensive and technically complex.

He then shows that in Group A the portion of agents using a new technology will follow a cumulative lognormal time path, because such innovations experience major improvements in their early years but thereafter there are fewer improvements. The number of agents using Group B innovations, in contrast, will follow a cumulative normal time path, because they experience improvement for many years after their first commercial introduction.

It is assumed that the rate of decline in *ER* (expected pay-off period) will differ between the two groups as follows:

$$d[(d\theta_1/dt)/\theta_1]/dt \quad \begin{cases} > 0 & \text{for Group A innovations,} \\ = 0 & \text{for Group B innovations.} \end{cases}$$

In other words, for Group A innovations, this reflects a slowing down over time in learning by manufacturers and in the returns from information search by potential adopters, but no such slowing down for the more expensive and complex group B innovations, at least over the diffusion period.

It also seems reasonable to suppose that the general shape of the time path for θ_2 will vary between Group A and Group B innovations:

$$d[(d\theta_2/dt)/\theta_2]/dt \quad \begin{cases} < 0 & \text{for Group A innovations,} \\ = 0 & \text{for Group B innovations.} \end{cases}$$

Thus it is assumed that the relaxation in R* (critical pay-off period) will level off over time for Group A innovations but will be more sustained for Group B innovations. Combining the conditions imposed on θ_1 and on θ_2, we have the following relations:

$$d[(d\theta/dt)/\theta]/dt \quad \begin{cases} < 0 & \text{for Group A innovations,} \\ = 0 & \text{for Group B innovations.} \end{cases}$$

Specifically, Davies [19] considers the following two types of functions:

$$\theta(t) = \alpha \cdot t^{\phi} \quad \text{for Group A innovations,}$$

and

$$\theta(t) = \alpha \cdot e^{\phi \cdot t} \quad \text{for Group B innovations,}$$

where $\alpha > 0$; $0 < \phi < 1$.

If we assume a specific functional form in the distribution of agent size, we can derive the diffusion path. Let agent size S be lognormally distributed, i.e. log S be normally distributed with mean μ_s and variance σ_s^2, and let $N(\cdot/\cdot,\cdot)$ be the normal distribution function, then, Davies [19] derives the following formula:

$$Q(t) = N(\log t/\mu_d, \sigma_d^2), \text{ for Group A;} \tag{4}$$

$$Q(t) = N(t/\mu_d, \sigma_d^2), \text{ for Group B,} \tag{5}$$

where,

$$\mu_d = -(\log \alpha + \beta \cdot \mu_s)/\phi;$$
$$\sigma_d^2 = (\sigma^2 + \beta^2 \cdot \sigma_s^2)/\phi^2.$$

Thus we derived the three types of diffusion path: a logistic curve for a psychological model, a cumulative lognormal time path for Group A innovations, and a cumulative normal time path for Group B innovations. The difference in time-shape among them can be seen in Figure 6.1. As can be seen in the figure, the technically complex and expensive Group B innovations should follow a symmetrical S-shaped diffusion curve not dissimilar to the logistic curve. The main difference is in the relatively longer tails of the logistic. For the less costly and relatively simple Group A innovations, however, the diffusion curve should take on a positively skewed shape, but will slow quite noticeably thereafter.

5.2 An alternative interpretation of diffusion models

In the light of techno-economic paradigm changes, we have to interpret diffusion models differently from conventional explanations and/or the authors of these models.

First of all, the logistic model should be understood in the light of techno-economic paradigm changes. An appropriate interpretation is that the process is epidemic, and that diffusion is constrained by the availability of information spreading mechanisms. Therefore, the decision is based on psychological pressure.

The probit approach, on the other hand, assumes that the decision is based on economic rather than psychological variables, i.e. an expectation of the pay-off. As far as the distinction between the two different time paths within the probit approach is concerned, however, technical simplicity is not necessarily correlated to organizational simplicity, because substantial organizational change might be needed to exploit new technology even if it is technically less sophisticated.

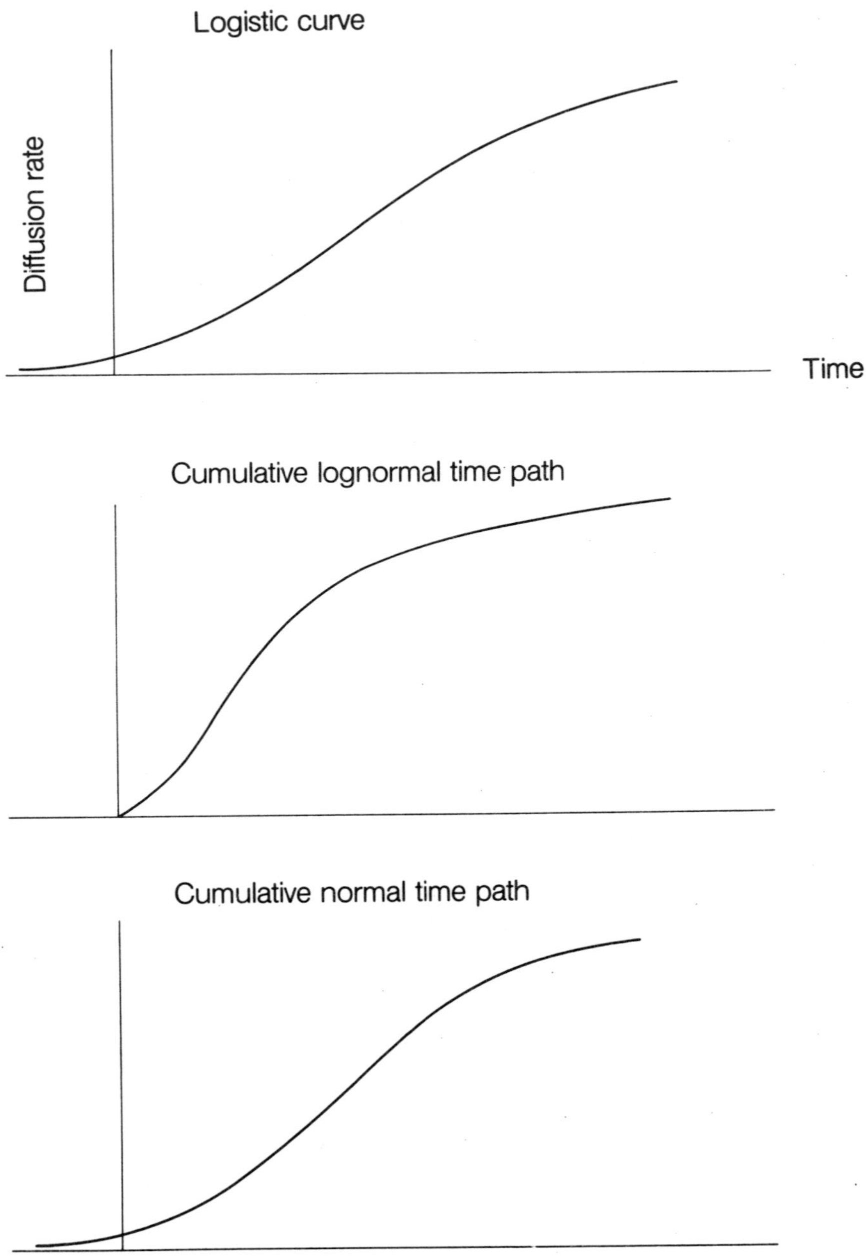

Figure 6.1 Three patterns of diffusion path

In fact, in the Davies mathematical formulation, the distinction does not necessarily come from technical complexity, but from whether the major improvements are concentrated in the early years or are scattered throughout the diffusion period. And the improvements might be either technical or organizational, i.e. they might come from organizational and institutional changes to accommodate the potential of the new innovations.

More specifically, we can interpret the Davies formulation in the following way. The agent's expectation of the pay-off period, $ER(t)$, will decline with time, because there will be successive improvements in the benefit/cost ratio. Because of the organizational inertia involved in adopting the techno-economic system, the benefit is not fully exploited without organizational change. And the cost of the labor displacement which is incurred by adopting the techno-economic system might be substantial. Innovation suppliers can gradually learn how to lessen institutional inertia by adding a certain function to the system. Most agents' views of the profitability of adoption can be improved by information resulting from search for how to implement the institutional changes associated with the adoption. Based on the above-described interpretations, we can assume that ER will decline monotonously with time.

The agent's critical pay-off period, $R^*(t)$, will increase with time, because of relaxations in the risk involved in implementing necessary organizational changes and the effect of competitive pressures which take the form of public criticism against not implementing these changes. Thus, we can also assume that R^* should increase monotonously with time.

We can distinguish the two groups of innovations in the following way: Group A innovations are those that can be implemented without substantial organizational change and are not necessarily simple technically; Group B innovations are those that can be implemented only with organizational change of a far-reaching character and might be unsophisticated technically.

For Group A innovations, the rate of decline in ER is attributed to a slow down over time in the availability of easier technical fixings by manufacturers to accommodate institutional inertia and in returns from information searches concerning easier organizational fixings by potential adopters. For Group B innovations, however, such a slowing down is not plausible, because neither easier technical nor organizational fixings are available or effective. The relaxation in R^* will level off over time for Group A, because of relatively lower risks involved in organizational changes, while it will be more sustained for Group B innovations because of the higher risks involved.

For the analysis of the techno-economic paradigm, therefore, we can interpret the Davies model in the following way: if the innovation requires a full-scale reorganization, then it follows the cumulative normal time path, because major improvements in productivity are possible only after a period of organizational change. On the other hand, if the innovation does not need organizational change, it will follow the cumulative lognormal time path, because its potential can be easily exploited in the early years.

6. Quantitative evidence: diffusion of computer into prefectural governments

As described before, the diffusion of computer utilization in prefectural government was selected for study. [20] Quantitative evidence of a new techno-economic system is presented in the form of statistical tests. And it will be shown that the quantitative study can identify the changes in a techno-economic paradigm. [21]

6.1 Test for shape of size distribution

Since the Davies model is heavily dependent upon the lognormality of size distribution, we first have to make a statistical test as to whether the size distribution of the prefectural governments is in fact lognormal or not. The population is selected as the indicator of size, because other indicators such as budget are found to be highly correlated to population size.

The data of population distribution in the years of 1960, 1965, 1970, 1975, 1980 were collected and tested. The distribution curves are obtained by dividing the logarithmic transformation of population into five equal intervals. Then, the *X*-square test is applied to these data. The results of the test are shown in Table 6.11. Since we are interested only in the fitness of the distribution curve, a degree of freedom is four.

Table 6.11 Statistical test for size distribution

Year	X^2-value
1960	4.11
1965	5.24
1970	6.51
1975	7.33
1980	5.43
5% point	9.49

Since the 5 percent point of X^2-distribution on 4 degrees of freedom is 9.49, as is shown in the table, the null hypothesis is not rejected in all the years. That is, one cannot deny that the population distribution follows the lognormal distribution. On the basis of the above results, we can assume that the size distribution of the prefectural governments does follow the lognormal distribution.

6.2 Grouping of task categories

We then come to the test of fitness. As described before, the dates of first installment of a computer are available every year since 1963, although saturation is reached in 1978. As far as the utilization of the computer is concerned, the dates of first utilization of the installed computer to perform each of the forty-one specified task categories are also available every year since 1963. And the objective of our test is to determine which best describes data: the logistic curve; the cumulative lognormal curve; or the cumulative normal curve.

The nine task categories which are shown in Table 6.10, are the ones in which the diffusion rate is relatively high by the year 1985. However, there are the two categories in which the available data seem to be less reliable because of an obvious inconsistency with present analysis in the data of 1976. These two categories are payroll calculation and forest plan, in which the diffusion rate peaked as high as 94 percent in 1976, and decreased drastically afterwards. On the basis of the data reliability, therefore, the following seven task categories were selected for the statistical test:

1. personnel management,
2. taxation for automobiles,
3. taxation for corporations,
4. management of issuing bonds,
5. production of the designated statistics,
6. forecasting and planning, and
7. cost estimation of civil works.

We are interested in obtaining evidence which indicates that the barriers for diffusion of the techno-economic system lie more in institutional inertia than in the system's technical complexity. Therefore, we have to identify the organizational complexity and technical complexity of these seven task categories. In order to accentuate the importance of organizational aspects, it is ideal to have the following two groups: one where the computerization of a task category is technically simple but organizationally complex, and the one where it is technically complex but organizationally easier.

On the basis of the considerations described above, we have the following two groupings of the task categories. The first group, 1–5, is characterized by routine work, mechanistic processing and technical simplicity, but involves large-scale data-processing, operation in a centralized mode, interdependence upon other agents, reliance on communication network and a high level of necessity of standardization. The second group, 6–7, is characterized by unstructured work, organic processing, and technical sophistication, but involves only small-scale, decentralized, independent, local-use mode operation, with less standardization needed.

Regarding technical sophistication, the computerization of the second group of task categories must go along with the development of a new science such as planning science, while that of the first group can be realized by solving the technical problems of data-processing. In other words, sophisticated software has to be developed to realize effective computerization of the second group of task categories, whereas incremental improvement in hardware might make possible the computerization of the first group of task categories. Moreover, the utilization of the computer for the second group requires particular expertise specific to each field of planning, while the first group primarily involves employing the system for clerical work. Therefore, we can conclude that the second group is technically sophisticated, while the first group is technically simple.

On the basis of the above consideration, we can thus summarize by noting that the introduction of computers into the first group of tasks will require changes that are complex organizationally but simple technically, and that the second group includes those that are sophisticated technically but simple organizationally.

6.3 Method for statistical fit

For the logistic curve of equation (1), empirical tests are fairly straightforward using regression analysis on the linear transformation:

$$\log\{m_t/(n-m_t)\} = \alpha + \beta\cdot t.$$

For estimation of the probit approach, both forms of (4) and (5) can be linearized by the following transformation:

$$Q(t) = N(z_t|0,1).$$

Given values for $Q(t)$, z_t can be read off from standard normal tables. Rearranging equations (4) and (5) in terms of the standard normal function, it follows that:

$$z_t \begin{cases} = (\log t - \mu_d)/\sigma_d & \text{for cumulative lognormal,} \\ = (t - \mu_d)/\sigma_d & \text{for cumulative normal.} \end{cases}$$

Therefore, we can estimate the coefficients from the following regression:

$$z_t \begin{cases} = a_1 + b_1\cdot\log t & \text{for cumulative lognormal,} \\ = a_2 + b_2\cdot t & \text{for cumulative normal,} \end{cases}$$

where z_t refers to the normal equivalent deviate of the level of diffusion (m_t/n) in year t. The regression estimates of a and b can be interpreted as:

$$a = -\mu_d/\sigma_d, \qquad b = 1/\sigma_d.$$

7. Identification of diffusion pattern

7.1 Test results

Regression analysis is first applied for the diffusion of the computer utilization for the seven specified task categories. For example, regression analysis applied for the task category of personnel management, yields the following results with the coefficient of determination R^2 and t-values in the parenthesis:

(a) for logistic curve,

$$\log\{m_t/(n-m_t)\} = \underset{(-32.47)}{-7.0701} + \underset{(10.44)}{0.1630 \cdot t}, \quad R^2 = 0.858,$$

(b) for lognormal time path,

$$z_t = \underset{(-7.626)}{-2.850} + \underset{(8.040)}{1.216 \cdot \log t}, \quad R^2 = 0.844,$$

(c) for normal time path,

$$z_t = \underset{(-14.04)}{-2.016} + \underset{(14.98)}{0.154 \cdot t}, \quad R^2 = 0.926.$$

This test result shows that all the T-value tests cleared the significant test. That is, the null hypothesis concerning all the parameters can be rejected at significance level of 0.05. However, the coefficient of determination is the highest in the cumulative normal time path, and this is the only time path in which the coefficient of determination is above 0.9. Therefore, we can fairly clearly say that the cumulative normal time path is identified as the diffusion path of the computer utilization for the task category of personnel management.

For the other six categories, the test results show that there is at least one time path in which the coefficient of determination is above 0.9, and all the t-value tests cleared the significance tests at the level of 0.05.

Our purpose is then to identify the diffusion pattern of its utilization for the seven task categories. In other words, the three types of time function, that is, the logistic curve, cumulative lognormal time path, and cumulative normal time path are compared in terms of R^2 and its t-value. The results are shown in the Table 6.12 with t-values in the parenthesis.

As shown in the table, none of the seven task categories of computer utilization is identified as following the logistic curve. That is, decision-making about the utilization of computers is based on an expectation and speculation of the pay-off periods, and not on the psychological and epidemic behavior. Therefore, as far as the utilization of computers is concerned, the identification problem is centered around the comparison between the cumulative lognormal path and the cumulative normal path.

Table 6.12 Comparison among three patterns

Task category	Logistic curve	Cumulative lognormal	Cumulative normal
personnel management	0.858 (10.44)	0.844 (9.865)	**0.926** (14.98)
taxation for automobiles	0.774 (8.265)	**0.927** (15.90)	0.900 (13.40)
taxation for corporations	0.930 *(15.50)*	0.929 *(15.39)*	**0.990** *(41.50)*
management of issuing bonds	*0.898* (12.61)	*0.740* (7.154)	**0.941** (16.96)
production of statistics	0.884 (11.69)	0.903 (12.92)	**0.971** (24.51)
forecasting & planning	0.713 (6.298)	**0.938** (15.51)	0.843 (9.270)
cost estimation of civil works	0.795 (6.524)	**0.940** (13.17)	0.905 (10.25)

On the other hand, when it comes to the diffusion of the installation of the computer without any specified task category, regression analysis yields the following results:

(a) for logistic curve,

$$\log\{m_t/(n-m_t)\} = -4.3281 + 0.5279 \cdot t, \quad R^2 = 0.986,$$
$$\qquad (-26.91) \quad (29.84)$$

(b) for lognormal time path,

$$z_t = -2.940 + 1.544 \cdot \log t, \quad R^2 = 0.833,$$
$$\qquad (-7.626) \quad (8.040)$$

(c) for normal time path,

$$z_t = -2.417 + 0.294 \cdot t, \quad R^2 = 0.985.$$
$$\qquad (-26.00) \quad (28.71)$$

On the basis of the above results, the logistic curve is identified as the diffusion path that best describes the diffusion of the installation of the computer. First of all, the lognormal time path is excluded from the possible candidates, because it yields the lowest values in terms of both R^2 and of t-value. Although the R^2 difference between the logistic curve and the cumulative normal time path is minimal, the logistic curve also has the higher t-value. However, considering the fact that none of the task categories of computer utilization is identified as following the logistic curve, we can positively identify the logistic curve as the diffusion path of the computer installation.

Therefore, we can conclude that the installation behavior is in fact psychological and epidemic. In other words, the decision-making about the computer installation is based on psychological and epidemic pressures with less rational and economic thinking. Therefore, the speed of the diffusion is determined by that of information spreading.

7.2 Distinction between the two groups

As depicted in Table 6.12, we come to a fairly clear-cut distinction between the two groups of task categories. Those task categories belonging to the first group are identified as the cumulative normal time path, with one exception, and those belonging to second group are identified as the cumulative lognormal path.

Thus we can say that there is only one obvious exception: the taxation for automobiles is identified as the cumulative lognormal path, although it was assumed to belong to the first group. Having this result, we can think of a unique feature of this task category. In Japan, car ownership increased rapidly during the 1970s, compared to the other growth which corresponds to the other task categories, as shown in Table 6.13. In this table, the growth in the number of passenger cars registered is compared to those in the number of prefectural government employees and the number of corporations, which correspond to task categories of personnel management and of taxation for corporations respectively.

As seen in the table, the area for the task category of the taxation for automobiles has expanded substantially compared to the other areas of other task categories. Therefore, we can conclude that in an expanding area, the introduction of computers does not threaten anybody with unemployment, and if employees' workloads are growing because hiring is not keeping up with increased work, then computers will make their lives easier and reduce their workload. Thus there was relatively less institutional inertia against diffusion of computers for this task. Once technical problems were solved, diffusion was faster, i.e. there were fewer institutional problems to be solved.

7.3 Derivation of taxonomy

On the basis of the statistical results mentioned above, we can develop a taxonomy of diffusion patterns as shown in Table 6.14. As depicted in the table, the information-spreading mechanism determines the rate of computer installation. Evidence that institutional inertia rather than technical sophistication determines diffusion patterns is two-fold.

First, the comparison between the first group and the second group yields the following conclusion: the diffusion of computer utilization is slow, that is, it follows the cumulative normal time path, when it is among the first group of task categories, in which

Table 6.13 Growth index of car registration compared to other growth index (index: 1963=1.00)

Year	Number of employees in prefectural governments	Number of passenger cars registered	Number of corporations	Budget size in prefectural governments
1963	1.00	1.00	1.00	1.00
1964	1.09	1.35	1.03	1.13
1965	1.11	1.74	1.11	1.28
1966	1.12	2.29	1.20	1.43
1967	1.14	3.01	1.26	1.60
1968	1.15	3.98	1.68	1.87
1969	1.17	5.11	1.77	2.16
1970	1.19	6.28	1.88	2.52
1971	1.22	7.57	1.98	3.02
1972	1.24	9.24	2.06	3.65
1973	1.28	10.75	2.22	3.63
1974	1.31	12.24	2.38	3.60
1975	1.31	13.74	2.60	3.89
1976	1.33	15.02	2.78	4.12
1977	1.34	16.28	2.91	4.65
1978	1.37	17.78	3.07	5.43
1979	1.39	19.05	3.25	5.52
1980	1.41	19.97	3.37	5.04
1981	1.42	20.87	3.69	5.33
1982	1.43	21.68	3.76	5.45
1983	1.44	22.51	3.86	5.72
1984	1.44	23.19	3.91	5.90
1985	1.44	23.96	3.94	6.27
1986	1.44	24.73	4.03	7.21
1987	1.44	25.79	4.15	8.19

Source: *Japanese Statistics*, Statistical Bureau, Prime Minister's office, Government of Japan.

organizational change is necessary, even if it is simple technically. On the other hand, diffusion is rapid, follows the cumulative lognormal time path, when it is among the second group, in which no substantial organizational change is needed, even if it is sophisticated technically. Thus we can conclude that the diffusion of computer utilization is in fact a function of institutional inertia rather than of technical change.

Secondly, within those task categories which belong to the first group, the diffusion of computers in the task category which is so expanding that there is no well-established institutional framework for performing this task is identified as following the cumulative lognormal time path. This represents a rapid diffusion pattern. Those task categories for which there

Table 6.14 Derived taxonomy of diffusion path

Category	Characterization of tasks			Identification		
	complexity of organization	technical sophistication	established institutional framework	pattern	decision-making	determinant of diffusion
installation of computer				logistic	psychological epidemic	information
utilization for: personnel management tax for corporation management of bonds produce statistics	large scale standardization centralised operation	simple	well-established	cumulative normal	expected pay-off	institution
utilization for: tax of automobiles	same as above	same as above	not yet established	cumulative lognormal	expected pay-off	institution or technology
utilization for: forecast & planning civil work estimate	simple	organic processing unstructured work sophistication	fairly established	cumulative lognormal	expected pay-off	technology

was an established institutional framework are identified as following the cumulative normal time path, which represents a slow diffusion pattern.

8. Conclusion

Through empirical and quantitative studies of diffusion of new technologies, we obtained new findings and clarified some of the effects of institutional and organizational influences on techno-economic paradigm change.

Whereas it has often been argued that institutional factors act as a barrier to the diffusion of technologies, this chapter has shown that this is not always the case, and that in some instances the effect can be exactly the opposite, with institutional response accelerating diffusion. The analysis presented in this chapter also has shown that within institutions it is important to consider certain organizational dichotomies which can act as significant determinants of diffusion. Significant dichotomies were found between the sophistication and simplicity of the technology and of the organization, and at a second level, between new and existing activities in the organization.

In the case of facsimile communication, widespread diffusion in Japan began only after the revision of the Public Telecommunication Law. In this sense, regulation prevented the potential of facsimile communication from being fully exploited. However, once the change in regulation was realized, facsimile communication diffused very rapidly. However, the rapid world wide diffusion of facsimiles was not a result of regulation, because the regulation was different abroad from that in Japan. It indicates that the facsimile turned out not only to be a replacement of its predecessors such as telex communication, but to be dominant over them in every aspect of its functions. In other words, the facsimile overrode its predecessors. This was the reason why it diffused so rapidly in these countries which have different cultural traditions and communicate in different languages.

The case study on the diffusion of CAT scanners in Japan alludes to the fact that the relevancy of this technology to institutional changes becomes an accelerator rather than a barrier. If a technology is very novel to users and has an intrinsic usefulness to them, as occurred in CAT scanners with doctors, the introduction of the new technology triggers a chain reaction of institutional changes which creates a favorable environment for the rapid diffusion of the new technology.

The quantitative analysis of the diffusion of the computer in Japanese prefectural governments reveals that its diffusion process is not as simple as described so far in terms of techno-economic paradigm changes. First, a distinction between the installation of the computer and its utilization must be made. The installation behavior follows conventional wisdom: the psychological approach is valid, the diffusion phenomena is epidemic, and its time path follows the logistic curve which is the diffusion pattern of the conventional technology. However, when it comes to the utilization behavior, the probit approach is valid, and the theory of techno-economic paradigm change holds: the institutional inertia is a determinant of diffusion rather than technological sophistication.

However, we need several reservations about the simple argument posed by the techno-economic paradigm change. Our study yields two levels of dichotomy. The first dichotomy is one between organizational complexity and technical complexity. Organizational inertia is in fact a determinant of the diffusion of computer utilization when it is applied to those activities which are complex organizationally. However, organizational inertia is not a determinant of diffusion when the computer is utilized for those activities which are less complex organizationally but which require technical sophistication.

The other dichotomy is whether a new technology is applied to old activities or to new activities. Institutional inertia is a determinant when the computer is applied to conventional activities for which the institutional framework is already well established. On the other hand, institutional inertia is not critical to the diffusion of the computer when it is applied to a new activity for which the institutional framework is not yet fully developed, i.e. when the institutional framework is to be developed in parallel with the diffusion of the new technology.

References

[1] C. Freeman, *Technology Policy and Economic Performance*, pp. 60–79, Pinter Publishers, 1987.

[2] G. Dosi, C. Freeman, R. Nelson, G. Silverberg, and L. Soete, *Technical Change and Economic Theory*, Pinter Publishers, 1988.

[3] F. Kodama, 'Paradigm shift in technology diffusion: from technical change to institutional inertia', presented at the *Stockholm Conference on Technology Management and International Business*, Institute for Management of Innovation and Technology, Stockholm, 1990.

[4] F. Kodama, 'Demand articulation: targeted technology development', in *Technik-Politik: Angeschichts der Umwelt-Katastrophe*, Ed. H. Krupp, pp. 273–294, Physica-Verlag Heidelberg, 1990.

[5] M. Trajtenberg, *Economic Analysis of Product Innovation: The case of CT scanners'*, pp.54–8, Harvard University Press, 1990.

[6] J.S. Metcalfe, 'The diffusion of innovation: an interpretative survey', in *Technical Change and Economic Theory*, G. Dosi, C. Freeman, R. Nelson, G. Silverberg and L. Soete, Pinter Publishers, 1988, p. 561.

[7] Ministry of Home Affairs, *Annual Report on Computers in Local Governments* (Tokyo, 1985).

[8] Management and Coordination Agency, *Annual Report on Utilization of Computers in Government* (Tokyo, 1985).

[9] P. Stoneman, *Technological Diffusion and the Computer Revolution, The U.K. Experience*, Cambridge University Press, 1976.

[10] P. Stoneman, *The Economic Analysis of Technological Change*, Oxford University Press, 1983.

[11] E. Mansfield, *Industrial Research and Technological Innovation*, W.W. Norton, 1968.

[12] L. Nabseth and G.F. Ray, *The Diffusion of New Industrial Processes: An international study*, Cambridge University Press, 1974.
[13] E. Rogers, *Diffusion of Innovations*, Free Press of Glencoe, 1962.
[14] V. Peterka, *Macro-dynamics of technological change: market penetration by new technologies*, PR-77-22, International Institute for Applied Systems Analysis, Laxenburg, Austria, November 1977.
[15] J.C. Fisher and R.H. Pry, 'A simple substitution model of technological change', *Technological Forecasting and Social Change*, vol. 3, pp. 75–88, 1971.
[16] A.W. Blackman, 'The market dynamics of technological substitution', *Technological Forecasting and Social Change*, vol. 6, pp. 41–63, 1974.
[17] P.A. David, *A Contribution to the Theory of Diffusion*, Stanford Center for Research in Economic Growth, Memorandum no. 71, Stanford University, 1969.
[18] P.A. David, *Technical Choice, Innovation and Economic Growth*, Cambridge University Press, 1975.
[19] S. Davies, *The Diffusion of Process Innovations*, Cambridge University Press, 1979.
[20] T. Muraoka, 'Analysis on diffusion of innovation', MS dissertation submitted to Graduate School of Policy Science, Saitama University, 1986.
[21] F. Kodama, 'Can empirical quantitative study identify changes in techno-economic paradigm?', in *Science, Technology and Industry Review* no. 7, pp. 101–129, OECD, Paris, July 1990.

Epilogue

The distinguishing characteristic of science and technology (S&T) policy research is that it is necessary, especially for frontrunning nations. For example, economics policy research is just as important for developing nations as for advanced countries and there is an academic field of 'development economics'. On the contrary, science and technology policy research for developing countries is very difficult to format into one academic field, as shown by the difficulties experienced in implementing the ideal concept of 'appropriate technology' in developing nations. S&T policy research is also not quite as effective for nations who are catching up, as seen in the case of Japan in the recent past, where S&T policy research had not been active for the last twenty years.

After World War II, S&T policy research was born first in the United Kingdom, when Freeman's group established the Science Policy Research Unit (SPRU) at Sussex University. This was because at that time the United Kingdom was the frontrunner of the world in science and technology, and had a tradition of academic thinking. Centers of S&T policy research moved to the United States as the centers of science and technology moved. In the United States, S&T policy research was developed in several universities, diffusing into existing departments or programs. US S&T policy research dealt mainly with the management of big government projects such as nuclear development or space development.

Japan is becoming one of the frontrunners in industrial technology, which means that prominent S&T policy researchers all over the world now pay more attention to Japan. Considering this change more deeply, one can understand the reason for the researcher's academic interest: the paradigm of technological innovation is shifting.

The reason that S&T policy research after WWII began in the United Kingdom is that extensive technology development had taken place there since the Industrial Revolution. The research in SPRU often deals with a database of those cases; the most typical research statistically analyzes more than 2,000 cases which have accumulated for more than 100 years. When centers of S&T policy research subsequently moved to the United States, they focused on the many cases of government projects such as nuclear development or space development. It can be said that the methods of big project management such as PERT (program evaluation and review technique) or PATTERN (planning assistance through technical evaluation of relevance numbers) were initiated in that era.

Why then are centers growing in Japan? We cannot ignore the emergence of civilian high technologies. The innovation process of high technology seems to be different from that for conventional technologies, and therefore cases of recent technology development in Japan should be analyzed. In other words, analysis only of cases in Western Europe or the United States does not lead to a new scientific framework for analyzing innovation of high technologies. It cannot be denied that this recognition, whether they were aware of it or not, increased prominent researchers' scientific interest in Japan.

Considering this historical background, the principles for carrying out S&T policy research in Japan should become clear. *First*, we should remember that many examples of technology development in this high-tech era exist in Japan. And therefore we should conduct objective and scientific analysis of the technology innovation process of this era. One consensus among science policy researchers is the inadequacy of the linear model in explaining the parameters of innovation. Researchers all over the world think the traditional innovation model needs overall reconsideration.

The most effective approach to counter-argue the linear model is to choose cases of Japanese high-tech development, because Japan succeeded in developing high technology in a social and cultural context different from those in Europe and United States. This does not mean that this research should be done only by researchers in Japan; the experiences should be common property and shared with researchers everywhere. In this way Japan can contribute to the world.

Second, S&T policy research will not progress simply by only studying Japanese cases. Collecting cases and constructing a database will not contribute to intellectual progress in the field but will only repeat what has been done before. The US contribution of new mathematical methods and the analysis of government management of big R&D projects in S&T policy research should be referenced, that is, the fusion of new objects and new methods would make a new field of study. The Japanese contribution should aim at the scientific method to be applied to areas other than management, for example to governmental and corporate policies. In short, it should apply scientific and quantitative method to the analysis of the innovation process of high technology which is dynamic and nonlinear.

Third, statistical data have become abundant. This means that the environment for applying mathematical methods to the quantitative analysis of S&T policy is improving. Japan has always been one of the best accumulators of chronological data, which are indispensable in the quantification of S&T activities.

As far as the interaction between policy-makers and policy researchers is concerned, it has been thought that policy-makers formulate problems and present the alternatives to researchers, whereas researchers evaluate the alternatives by analyzing scientifically and then make a recommendation to policy-makers. But policy-makers do not seem to be able to formulate problems except for political issues. On the other hand, policy analysts are not only interested in solving the problems within the alternatives. Policy-makers are annoyed when they find that any existing policy alternative is not effective, and hope for analysts' help. Policy analysts, who are scientists of academic inclination, are only interested in solving complicated puzzles.

In other words, what is a dilemma for policy-makers is a matter of interest for academics. Considering this relationship, each side can devote their efforts to their own interest arena, and the problem annoying policy-makers will surely be solved.

I should like to conclude this book with a small essay on Japanese information technology, which is becoming a driving force behind the global wave of political change. The collapse of one communist regime after another in Eastern Europe over the past several months will mark an important chapter in world history.

If one were to examine recent successes in democratization from the perspective of technology, it could be concluded that Japanese information technology has played a significant behind-the-scenes role. This hypothesis can be tested by examining the role of Japanese cassette tape recorders in the Iranian revolution, the role of Japanese video cassette recorders in the Philippines' 'people power' revolution and the role of facsimiles in the recent Chinese pro-democracy movement which ended in the military's violent crackdown in Tiananmen Square. Although Japan developed these technologies for its own benefit, they were used in unanticipated ways and became an extremely powerful means for toppling oppressive regimes.

First, at the time of the Iranian revolution, the voice of Khomeini was equivalent to the voice of God. The people had to hear it directly. If the cassette tape recorder had not been in existence, however, this would not have been possible. Of course, Khomeini's voice could have been heard through broadcasts. The Shah's government, however, would undoubtedly have intervened to thwart such broadcasts. The key point is that Khomeini's actual voice could be passed along and listened to by very small groups. This underscores the effectiveness of the cassette tape recorder.

The Philippine revolution, for its part, was a revolution made possible by the VCR. In spite of suppression by the Marcos administration, video footage filmed overseas could be smuggled into the Philippines. This in turn made it impossible for the Marcos regime to hide the truth. Moreover, the live broadcasts seen around the world at the time of Marcos's fall are still fresh in our memories.

Lastly, the recent Tiananmen incident could be referred to as a facsimile-based revolution. As is widely known, Chinese civilization is based on written characters. There was no means of transmitting written messages until the advent of the facsimile. Moreover, facsimiles are unique in that it is difficult to preserve the confidentiality of communications. The combination of these two factors is what produced the Tiananmen incident. In other words, due to the progress in globalization, the situation became such that Chinese from all over the globe were sending a steady stream of information to facsimile machines, which were installed only in hotels. As messages were sent in the form of written characters, without the cover of confidentiality, there was ample opportunity for hotel employees to look at them before they were delivered. This opened a route through which information could be leaked to the public.

In the future, events in Eastern Europe will be of great importance, particularly with regard to East–West relations. During a visit to West Berlin last December, at the height of the drive for democratization, I had a chance to observe East Germans shopping in West Berlin with the 100 marks they had received from the West German government. The most popular items among shoppers from the East were bananas, followed by inexpensive Walkmans. Walkman technology was originally developed so that the user could enjoy music while riding on the train or, as the name suggests, while walking. But when considering the context of Eastern Europe countries under the thumb of repression, it becomes possible to imagine that a Walkman would be used as a means to listen to information without fear of eavesdroppers.

In this way, Japanese information technology could, in a sense, be viewed as a means to break down the political structure of oppression symbolized by the East–West conflict, and to construct something new in its place. In other words, the popularization of technology ultimately contributes to democratization. The point of the above is that the popularization of technology, regardless of how it is used, ultimately provides the people with a means to fight against oppression. On that point, 'popularized technology' can be seen as a technological concept to resolve East–West conflict.

Finally, from the perspective of technological history, an analysis of the correlation between technological systems and the political system of communism yields the following historical paradox. Specifically, the planned economy advocated by Karl Marx is predicated upon a concentration of all information in the hands of the central government. The technological means to realize this ideal, however, did not exist. Moreover, the advent of the computer, which could most thoroughly satisfy this technological need for the concentration of information, was brought about by the United States, which embodies the antithesis of the planned economy. In other words, there is a historical paradox in which a political system, built on the principle that it is impossible for all information to be concentrated within the central government, produced the technological means which would make possible the realization of the communist ideal.

Index